ETHNOBOTANY OF BULDHANA DISTRICT
Maharashtra

ETHNOBOTANY OF BULDHANA DISTRICT

Maharashtra

Dr. D.A. Patil
Associate Professor
Post-Graduate Department of Botany
S.S.V.P. Sanstha's L.K. Dr. P.R.Ghogrey
Science College, Dhule – 424 005
(Maharashtra)
&
P.S. Patil, Y.A. Dushing
U.P. Aher, Y.A. Ahirrao

2011
DAYA PUBLISHING HOUSE
Delhi - 110 035

ISBN 81-7035-674-1
ISBN 978-81-7035-674-5

Published by : **Daya Publishing House**
1123/74, Deva Ram Park
Tri Nagar, Delhi - 110 035
Phone: 27383999
Fax: (011) 23260116
e-mail : dayabooks@vsnl.com
website : www.dayabooks.com

Showroom : 4760-61/23, Ansari Road, Darya Ganj,
New Delhi - 110 002
Phone: 23245578, 23244987

Laser Typesetting : **Classic Computer Services**
Delhi - 110 035

Printed at : **Chawla Offset Printers**
Delhi - 110 052

PRINTED IN INDIA

Dedicated
To
The People of Buldhana District
(Maharashtra)
for sharing their wisdom with us

Preface

Indian plantlore was documented first by Garcia da Orta from Goa in 1563 in his 'Colloquios dos simples e drogas e cousas medicinals da India.' Christobel Acosta (1578) then published 'Tracado de las drogas medicinas de las Indians Orientalis'. The subsequent monumental work by Hendrik Adriaan Van Rheede was published in 12 volumes (1678-1693) of 'Hortus Malabaricus' (*cf.* Manilal,1999). The traditional plant based uses in India are also recorded by Roxburgh (1832) and Watt(1889-1896,1908). Later, botanisation in Indian subcontinent led Sir J. D. Hooker to state that 'India has no flora as a separate entity but is an admixture of the flora from adjacent countries. Floristic studies during post-independence proved that 'India has a flora of its own'. Indian plantwealth has not received a desired attention from utility point of view. Ethnobotanical study in our country is only about six decade– old. Researches on this line still do not suffice to prompt that 'India has not only flora of its own but also has its own indigenous plantlore/ medicolore'. Rational attempts to study ethnobotany will help redeem the situation. This science is itself (erroneosly) looked down by many Indian workers. It needs recognition. Statement made by Dr.R.A.Mashelkar, D.G. CSIR, India,(2002) is, however, encouraging and noteworthy. He opined that 'In real life, particularly in the developing world, there is a whole parallel knowledge based system, which is generated by people who live in the laboratory of life, based on their empirical wisdom and experience'.

It appears pertinent to remind Linnaeus, a legendary botanist. He (1732) returned from his field studies with Sami people of the arctic Lappland. He entered the University town of Uppsala (Sweden) wearing Sami clothing. The implication was– Linnaeus, although a hightly educated Swede did not see the indigenous and unschooled Sami people as benighted but instead as being possessed of dignity and worthy of respect (*cf.* Cox, 2005). His behaviour guides us for reducing gap between indigenous people and ethnobotanists so as to establish meaningful rapport.

Recent upsurgence of interest in Ethnobotany has been driven for many reasons. Important among these are: i) the accelerated loss of indigenous plantlore has focussed attention on the need to document indigenous traditions before they are lost forever ii) many biodiversity rich nations are considering indigenous plantlore as part of their national heritage iii) natural product research is again on the rise iv) a number of new drugs derived from plants have awakened interest in using indigenous knowledge to guide drug discovery efforts v) western medicine has become more open to alternative approaches to healing. The awareness about biodiversity caused by the 'Earth Summit' (UNEP, 1992), side effects of synthetic drugs and increasing interest in green chemistry/medicine have been also largely responsible. Documentation of indigenous wisdom has become, of late, more essential due to the strictures of the laws pertaining to patents and the increased awareness about practices of biopiracy.

Our earlier botanical and ethnobotanical forays in some districts of northern part of Maharashtra have been in the pursuit of viewpoints expressed above (Patil, 2003; Kshirsagar and Patil, 2008; Patil and Patil, 2006; Pawar and Patil, 2008). Buldhana district is investigated floristically (Diwakar and Sharma, 2000) however, it remained practically untapped ethnobotanically. We, therefore, extended our efforts on this line as an exclusive topic of research during 2006-2009. This book embodies the results of our sincere efforts in the region.

We are thankful to the people of Buldhana district for sharing their secrets with us . We are grateful to authorities of S.S.V.P.Santha for the facilities rendered. We are indebted to Professor (Dr.) R. M.Pai, Ex-Head, Post-Graduate Department of Botany, Dr. Babasaheb Ambedkar Marathwada University, Aurangabad (M.S.) for constant encouragement and critical suggestions. We tender our thanks to the Librarians of Dr. Babasaheb Ambedkar Marathwada University, Aurangabad; Jaykar Library, University of Pune, Pune; and Agarkar Research Institute (M A C S), Pune,College of Agriculture, Dhule (M.S.) and L.K. Dr. P.R.Ghogrey Science College, Dhule (M.S.). We shall be happy to receive comments from the users of the book.

Authors

Contents

PART – I
INTRODUCTION

1

Introduction

Although the relationship between plant and man is age-old, it is recognized as a branch of science very late.It is Harshberger (1895, 1896) who coined the term 'ethnobotany'. In his opinion, 'it is the study of plants used by primitive and aboriginal people'. Subsequently, Jones (1941) defined it as 'the study of the interrelationship of primitive man and plants'. Faulks (1958) considered the subject of ethnobotany as the total relationships between man and vegetation. A decade later, he revised his concept and treated ethnobotany as a unit of an ecological study specializing in the interaction of man and the plant world. Schultes (1962) interpreted ethnobotany as usually 'the study of relationship which exists between people of primitive society and their environment'. Heiser (1985) defined ethnobotany as the study of plants in relation to people. Bye (1985) and Jain (1986) also gave their own definitions. Manilal (1989) opined that 'the term today has come to denote the entire realm of direct relationship between plants and man.' Till few years, interpretations differed greatly, but generally include total natural relationship of plants kingdom with man and also his pet animals.

The word 'tradition' means oral transmission or practice from one generation to next. Although man has tried to keep written records especially after invention of printing, some traditions also passed over generations by word of mouth. The importance and significance of traditional knowledge in recent times, whether oral or written, are being recognized worldwide.

Man has inherent capacity to discriminate and exploit the natural resources in his surrounding to his advantage. He has also ability to innovate for developing material civilization. Economic systems/thoughts are intertwined with natural wealth. Even in modern period, several human societies still continue to inhabit in and around forests. It is further to be noted that the human societies although settled in some organized habitations in rural and semi-rural areas, their traditions have not changed much over centuries. Such human societies are the real treasures of ancient traditions and cultural heritages. They have useful and beneficial elements of floral and faunal wealth in their vicinity. Their experience and knowledge were subjected to modifications and refinements by the successive generations which have been also integrated with their traditions or customs. Their knowledge systems have thus also become broad–based in the past.

The success of aboriginal, tribal, primitive or rural societies in understanding plant and properties of their underground parts and aerial parts like stem, leaves, flowers and fruits is simply a result of long and intimate association with the flora and their dependence on them. Since their knowledge is based on experimentation on human beings, it warrants a careful consideration. The tribal people and ethnic races throughout the world have developed their own cultures, customs, religious rites, taboos, legends, myths, foods and medicinal practices. Several wild and cultivated plants play very important and vital role among cultures. This interrelationship has evolved over generations of experience and practice.

It also appears pertinent to note erosive forces affecting their knowledge systems and even plant wealth particularly in modern times. We are facing the problems caused due to the forces of acculturation coupled with the forces of vegetational deterioration, developmental activities and modernization. All these accelerated the rate of extinction of traditional knowledge as well as our biota.

The importance of traditional knowledge about plants and increasing commercial pressure generated more interest in gathering traditional information and utilization of the ambient bioresources. Dr. R.A.Mashelkar, Ex D.G., CSIR, (2002) appropriately stated, we quote it in verbatim, "In real life, particularly in the developing world, there is a whole parallel knowledge based system, which is generated by people who live in the laboratory of life, based on their empirical wisdom and experience. There is an urgent need to preserve, protect and add value to it".

India is well known for the diversities-of the land mass with almost all the climatic zones, biodiversity and ethnic diversity. India has been also included in the list of mega diversity centers of the world. The tribal and rural populace of different ethnicity resides in various ecological nitches, besides the inheritors of the Vedic and Pre-Vedic cultures. It is, therefore, obvious that the spectrum of knowledge about the life and the skill of utilizing the bioresources vary among the Indian communities. As far as the state of Maharashtra is concerned, it also represents a similar scenario from the aforesaid viewpoints. People from nearly all religions, many castes and tribes inhabit the state as in other parts of India. It is predominantly rural and agrarian. They still depend traditionally on wild plants in their vicinity. There are 35 districts in Maharashtra. The state comprises mainly four regions *viz.*, Vidarbha, Desh, Marathwada and Konkan. Buldhana is one of the district of Vidarbha region.

A perusal of literature revealed that Buldhana district has not been scanned from viewpoint of traditional plantlore to date. It has remained totally unexplored. It was, therefore, felt worthwhile to undertake fairly extensive and intensive investigation as an exclusive topic of research in this district of Maharashtra.

2
Geography

Buldhana district is located between 19°51′ and 21°17′ north latitude and 75°57′ and 76°59′ east longitude with an area of 9745 sq. km in Maharashtra (Map 1). It has common boundary with Jalgaon district of Khandesh region of Maharashtra on the west. It is bounded by the Nimar district of Madhya Pradesh on the north and on the east by Akola and Amravati districts of Vidarbha region of Maharashtra. On the south, it is delimited by the Jalna, Parbhani and Hingoli districts of Maharashtra (Map 2).

The district is divided into thirteen tehsils *viz.*, Buldhana, Motala, Malkapur, Nandura, Jalgaon (Jamod), Sangrampur, Shegaon, Khamgaon, Chikhali, Mehkar, Lonar, Sindkhed Raja and Dewulgaon Raja (Map 3). The district headquarter is Buldhana. The district was part of the state of Madhya Pradesh and transferred in 1956 to the erstwhile Bombay state. The district since May 1960 forms part of the state of Maharashtra. It is one of the north-westernmost districts in Maharashtra.The name of district Buldhana is said to be a corrupt form of 'Bhil Thana' meaning the place of Bhils, a predominant tribe in the district and the adjacent ranges of Satpura mountain (*cf.* Anonymous, 1976).

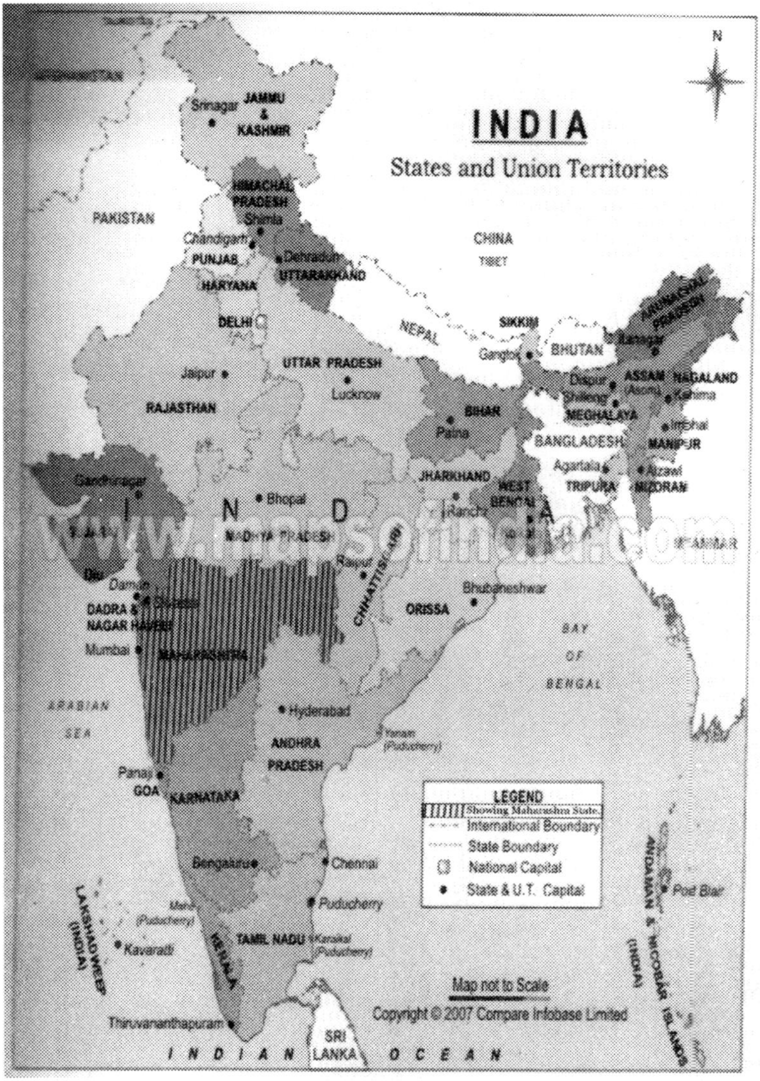

Map 1: Map of India Showing Maharastra State

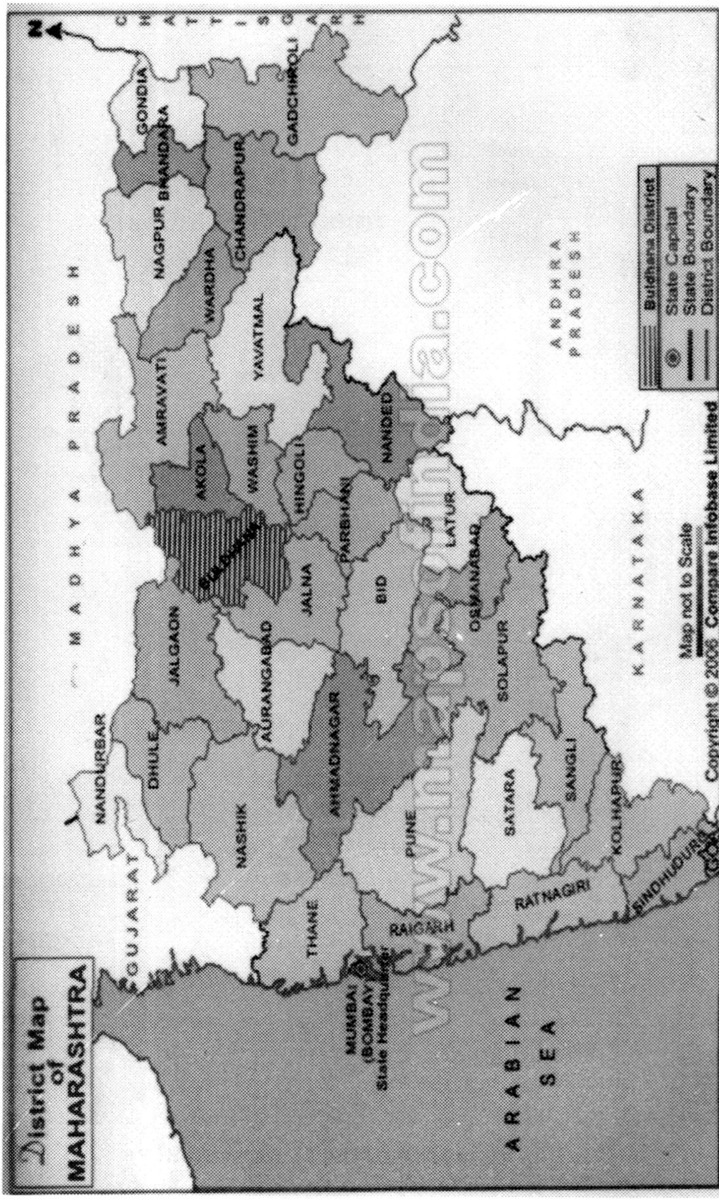

Map 2: Map of Maharashtra Shwoing Buldhana District

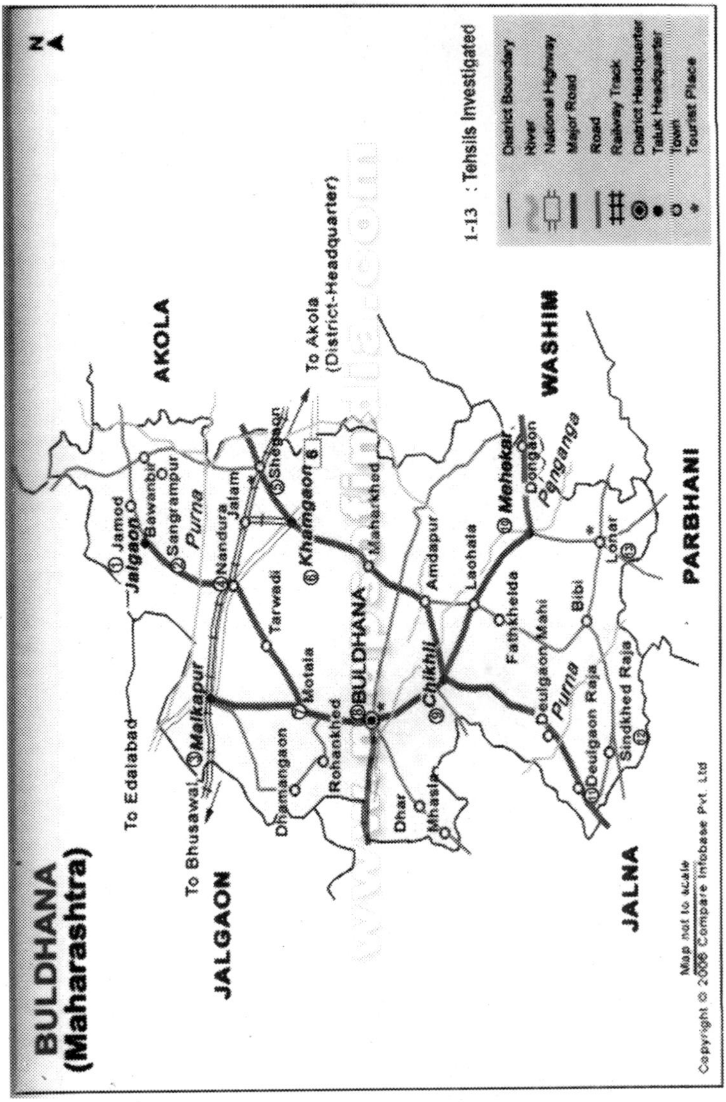

Map 3: Map of Buldhana District Showing Tehsils Under Investigation

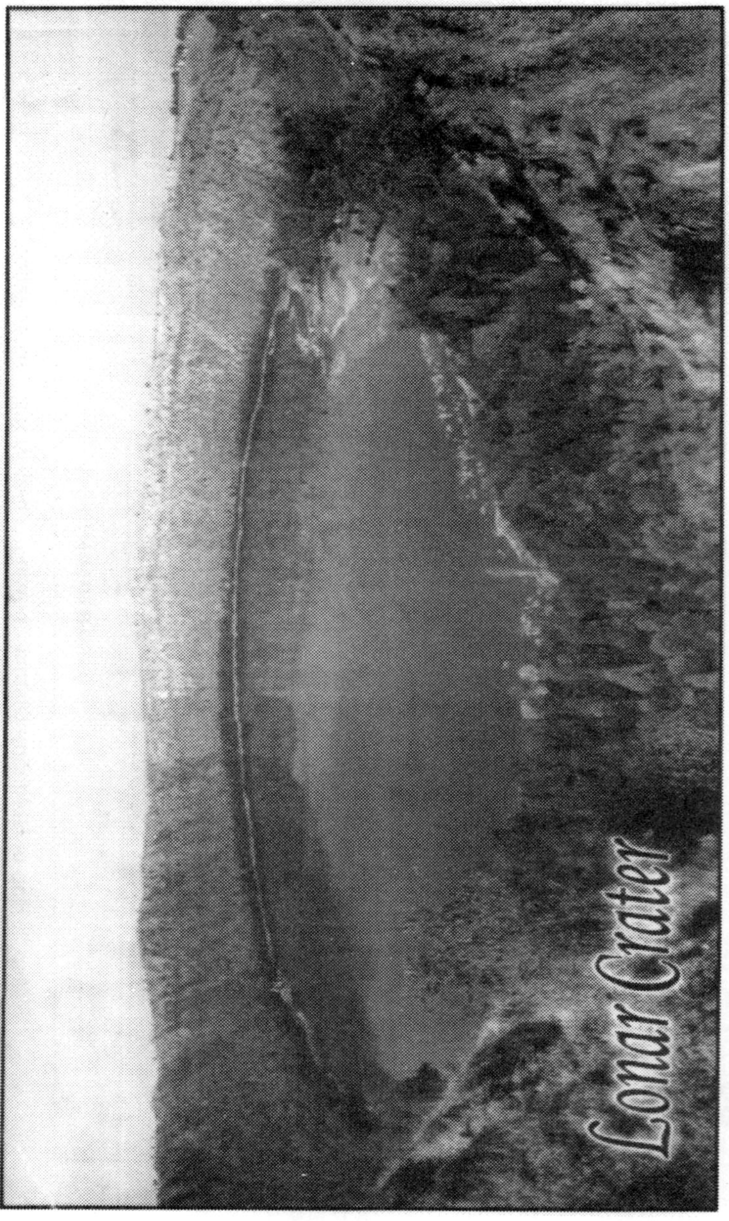

Plate 1: Land Cover

3
Geology and Soils

The major part of the district is occupied by rocks of the Deccan Volcanoes of Cretaceous-Eocene age, and a few alluvium patches of the Purna and Penganga basins. The trap rocks are generally fine to coarse-grained, dark grey to greenish-black basalts of vesicular and massive types. The hard compact massive flows are usually found on the hill tops like Melghat section. The soft and amygdular varieties are generally noticed on flanks of the hills or valley floors; weathering in the traps is characterized by spheroidal exfoliation. Apart from these, vertical, inclined, and columnar jointing are also observed in more massive types. At some places, the vesicular and non- vesicular flows are separated by thin beds of ash or scoriae. However, the typical inter-trappean sedimentary rocks are not found in this region. The amygdular varieties of flow contain secondary minerals such as zeolites, calcite and chalcedony. No dykes have been noted.

There is an extensive stretch of alluvium in the Purna valley between Paturda and the confluence of Purna River with that of Tapi River in adjacent Jalgaon district. At some places, a mixture of black cotton soil associated with sub-recent conglomeratic formation or light brown laterite material is observed. The alluvium of the plains is generally of greater depth, sometimes up to 50 meter e.g.

near Pimprala or Malkapur. Majority of the wells have brackish water.

The unique occurrence of picturesque 'Lonar Crater' in the great basaltic province of India is geologically interesting. It is a large isolated caldera-like depression. Its origin has been a subject of many investigations and discussion. It is situated about a kilometer to the south –west of the village Lonar with a diameter of 2000 meter and depth of 135 meter. The general elevation of its surrounding is 550 to 600 meter above msl (Plate 1). Its water is characterized by a high concentration of chlorides, carbonates and sodium with SiO_2, S_2, O_2, SO_2, Ca, Mg and K in minor amounts, besides traces of Boron.

4

Rivers and Lakes

The Penganga and Purna are main rivers in the district, apart from other small rivers or tributaries *viz.*, Katepurna, Nalganga, Vishwaganga, Gyanganga, Banganga, Dhamna, Koradi and the Mun. The Penganga rises close to the northern scrap edge of Buldhana plateau and Deulghat hills and enters the Akola district. It largely drains through the Chikhali and Mehkar tehsils. The Katepurna rises in the Ajanta ranges. It flows south easterly in Chikhali and Mehkar tehsils and enters Parbhani district. The Dhamna is a left blank tributary of the Katepurna. The Koradi also rises in Buldhana plateau and flows nearly parallel to the Penganga and joins it near Mehkar town. The Purna is the perennial river in the Payanghat plains. It rises in Gawilgarh hills of the adjacent district of Amaravati and joins the Tapi River in Jalgaon district. Banganga also rises similarly and joins the Purna River near Dewulgaon. The Mun River and its tributaries like the Torna, Vishwamitri and mass flow in Khamgaon tehsil and join the Purna. The Gyanganga rises in Buldhana plateau in the Matargaon forest area and joins the Purna near Yerli village after flowing in the Nandura tehsil. The Vishwaganga rises similarly and joins the Purna near Harsod, whereas Nalganga rises in the Dewulghat hills in the

Amdari forest area, flows through the Malakapur tehsil and joins the Purna near the village Narawel.

The Lonar Lake lies in Lonar tehsil in a nearly circular depression. The lake has circumference along the outer rim of about six km. The water is, however, brackish and saltish. The 'Skanda Purana' describes the lake representing a subterranean abode of 'Lonasura' who was unearthed by Lord Vishnu from his den. Geologists believe it formed during the closing phases of the Deccan trap eruptions, while some other think it an extinct crater of a volcano. There are other interpretations on record about its origin and formation (Plate 1).

5
Forests

There are reserved forests (about 1082 sq.km.) and protected forests (94.52 sq. km.) in Buldhana district. The forests are mainly situated in the Balaghat plateau of the Ajanta hills and in the Purna valley. They belong to the 'Southern Tropical Dry Deciduous' type. They are further categorized into seven types *viz.*, (1) Superior teak forest, (2) Inferior teak forest, (3) Anjan forest, (4) Babul forest (5) Mixed forest, (6) Salai forest and (7) Grasslands brushwood areas. The former two types are found in the Ambabarwa block, Mangeri, Ghatbori and Buldhana ranges. Mixed forests occur in all blocks. Teak, Anjan and Babul are the main trees and there are associates like species of *Terminalia, Aegle, Eugenia, Lagerstroemia, Anogeissus, Diospyros, Salmalia, Pterocarpus,* apart from bamboos.

6

Climate and Seasons

The district has hot summer, well distributed rainfall due to south-west monsoon season and dry winter from December to February. The hot months are from March to June. The average annual rainfall in this region is 796 mm. The north western part of Buldhana and Dhamangaon region receives comparatively more rainfall than the other parts of the district. The rainy monsoonal season lasts from June to September. October and November constitute the post-monsoonal months. The numbers of rainy days are usually 47 in a year. July is the rainiest month normally.

Temperature generally decreases till December. The coldest month begins in December. Mean daily temperature falls upto 15^{0}C in southern part of Buldhana district. It goes down to nearly three degrees above the freezing point of water. Temperature rises after February till May rapidly. May is the hottest month of the year. In southern part of district, the mean daily maximum temperature is about 38.3^0 C and the mean daily minimum 26.2^{0}C. In the northern part, it is 42.3^{0}C, while minimum is 27.4^{0}C. The day temperature rises up to 47^{0}C in the northern part and 42^{0}C in southern part of the district. The temperature appreciably drops with onset of the

Table 1: Average Rainfall in Last Ten Years (mm)

Tehsils	*1997*	*1998*	*1999*	*2000*	*2001*	*2002*	*2003*	*2004*	*2005*	*2006*
Buldhana	778.8	1063.0	684.6	563.0	734.8	1324.0	831.7	655.2	629.0	1224.0
Chikhali	675.0	983.0	687.3	615.4	752.2	789.8	635.9	662.2	832.0	1116.0
Dewulgaon Raja	471.0	1109.0	650.2	748.8	746.6	690.4	719.0	643.0	697.5	796.0
Mehkar	573.0	1088.0	1126.0	771.0	1065.0	876.0	779.0	599.8	647.0	1124.0
Lonar	543.0	1831.6	950.0	653.0	1185.0	1033.0	491.0	553.0	761.0	995.0
Sindkhed Raja	486.0	968.0	909.0	795.8	1287.0	1286.0	932.0	815.0	719.0	976.0
Malkapur	796.0	1314.4	489.0	571.0	524.0	1029.0	702.0	678.0	446.0	780.8
Motala	656.2	1065.9	644.8	626.0	689.0	970.0	891.0	541.0	345.0	984.0
Nandura	384.0	420.0	439.0	572.0	649.0	1075.4	496.0	465.0	376.0	660.0
Khamgaon	407.6	667.6	739.8	621.0	691.5	775.4	477.4	459.2	595.4	962.0
Shegaon	663.4	842.8	580.0	660.1	790.1	1007.0	956.7	378.6	591.9	959.8
Jalgaon (Jamod)	812.0	1134.4	511.0	558.0	553.6	1109.0	702.0	521.0	409.0	722.0
Sangrampur	750.0	761.0	487.0	520.0	549.5	905.0	922.0	380.0	543.0	930.0
Total	**7996.0**	**13252.7**	**8897.7**	**8275.1**	**10217.3**	**12960.1**	**8935.7**	**7351.0**	**7671.8**	**12319.6**
Average	**615.1**	**1019.4**	**684.4**	**636.5**	**785.9**	**996.9**	**687.4**	**565.5**	**590.1**	**947.7**

monsoon in second week of June. Secondary maximum day temperature is observed during October.

The relative humidity is high during the south-west monsoon season. The air is generally dry over the district. The relative humidities range between 25 to 30 per cent in the afternoons in the summer. The sky is cloudy heavily especially during the monsoon season, while in rest part of the year it is mostly clear or slightly clouded. Winds are light but increase in later part of the summer and in the monsoon season. During post-monsoon season, the wind blow from east or north-east. However, during the south-west monsoon season, winds blow between south-west and north-west. The post-monsoon storms and depressions from the Bay of Bengal affect the weather.

7
Plant Based Industries

The economic status of the people of Buldhana district can be studied from two aspects of economy: (i) industrial, (ii) agricultural. These also complement to each other. In this contribution, the agricultural and plant based industries are focused briefly. The people engaged in agriculture also suffer from disguised unemployment. This situation is redeemed by the development of small scale and cottage industries. The people of district are mainly agrarian. Apart from this, fairly rich forests are also observed in the district. The plants or their products are subjected to various treatments or conversions suitable to mankind. For the same sake, various industries are established. These are enlisted below:

(I) Agriculture Based Industries

(*i*) Industries for food-stuffs like rice, Atta, flour etc. by milling, dehusking and processing of food-grains

(*ii*) Industries for oil production *e.g.* edible oils like groundnut oil, sesame oil (*iii*) Industries for cotton spinning, ginning, pressing, baling, weaving etc.

(*iv*) Tobacco products

(*v*) Industries for Textile productions e.g. rope and cordage from plant fibres

(II) Forest Product Based Industries

(*i*) Industries for manufacture of bidi, brooms, etc.

(*ii*) Industries for wooden furniture, fixtures and structural goods like posts, beams, doors, windows etc.

(*iii*) Industries for manufacture of goods from bamboo, cane, gum etc.

(*iv*) Industries for production of medicines.

8

Crop Plants

Agriculture is the mainstay of the people of the district. Major population of the district depends on crops yielding cereals, pulses, vegetables, fruits, fibres, oils, sugar, spices and condiments, etc. Both the 'kharif' and 'rabi' crops are grown in the district. The staple food is comprised mainly of cereals and pulses. A resume of these is given below:

(*i*) Cereals and Millets: (a) *Sorghum bicolor* (Linn.) Moench. (b) *Pennisetum americanum* Schum. (c) *Triticum aestivum* L.(d) *Oryza sativa* Linn. (e) *Zea mays* L.

(*ii*) Pulses: (a) *Vigna mungo* (Linn.) Hepper (b) *Vigna aconitifolia* (Jacq.) Marechal (c) *Vigna radiata* (L.) Wilczek. var. *radiata* (d) *Vigna unguiculata* (L.) Walp. (e) *Cajanus cajan* (L.) Millsp. (f) *Cicer arientinum* L.

(*iii*) Vegetables: (a) *Solanum melongena* Linn. (b) *Lycopersicon lycopersicum* (L.) Sant. (c) *Cyamopsis tetragonoloba* (L.) Taub. (d) *Cucurbita maxima* Duch. (e) *Luff cylindrica* (L.) M.J. Roem. (f) *Luffa acutangula* Roxb.(g) *Coccinia grandis* (L.) Voigt. (h)*Momordica dioica* Roxb.(j) *Momordica charantia* L.(j) *Cucumis sativus* L. (k) *Abelmoschus esculentus* (L.) Moench.

(m)*Brassica oleracea* L. var. *capitata* L. (n) *Brassica oleracea* L. var. *botrytis* L. (o) *Moringa oleifera* Lam. (p)*Sesbania grandiflora* (L.) Pers. (q) *Spinacea oleracea* L. (r) *Lagenaria leucantha* (Roxb.) Rusby (s)*Daucus carota* L. (t) *Trigonella forenum-graecum* Linn.(u) *Raphanus sativus* L. and few others.

(*iv*) Edible Fruits: (a) *Musa paradisiaca* L. (b) *Psidium guajava* L. (c) *Mangifera indica* L. (d) *Carica papaya* L. (e) *Ziziphus mauritiana* Lamk. (f) *Citrus sinensis* L. (g) *Citrus aurantifolia* (Christm.) Sw. (h) *Phyllanthus emblica* L. (i) *Punica granatum* Linn. (j) *Citrullus lanatus* (Thunb.) Matsumura and Nakai (k) *Cucumis melo* L. (l) *Cucumis melo* L. var. utilissimus (Roxb.) Duthie and Fuller (m) *Annona squamosa* L. (n) *Annona reticulata* L. and few others.

(*v*) Spices and Condiments: (a) *Capsicum annuum* L. (b) *Coriandrum sativum* L. (c) *Allium cepa* L. (d) *Allium sativum* L. (e) *Murraya koenigii* (L.) Spr. (f) *Curcuma longa* L. and few others.

(*vi*) Edible Oil Yielding Plants: (a) *Arachis hypogaea* L. (b) *Sesamum indicum* Linn. (c) *Helianthus annuus* L. (d) *Carthamus tinctorius* L. (e) *Linum usitassimum* L. (f) *Glycine max* (L.) Merr.

(*vii*) Sugar Yielding Plants: *Saccharum officinarum* Linn.

(*viii*) Fibre Yielding Plants: (a) *Gossypium species* (b) *Hibiscus cannabinus* L. (c) *Agave vera- cruz* Mill. (d) *Crotalaria juncea* L.

(*ix*) Miscellaneous: (a) *Nicotiana tabacum* L. (b) *Piper betle* L.

9
History

Short prehistoric account is available of the Vidarbha region (*cf.* Anonymous, 1976). This region, including Buldhana district, was first occupied in the early Stone Age. Tools of trap rock like cleavers, scrapers and hand–axes have been found in the Vainganga valley. Neolithic Stone Age implements have been found at Kundinpur of this region. Orange–coloured pottery painted with red or black bands were associated. River valleys must have been the earliest habitations. They wore cotton garments. Economically, the people were in a pastoral-cum-hunting-cum-agricultural stage.

Legends have been recorded in the 'Puranas' particularly about the places like Lonar and Mehkar. The town of Lonar is believed founded in the 'Kritayuga'. It is mentioned in the 'Skandpurana' that the giant Lavanasura lived in a subterranean abode. From this place, he used to aspire waging war against the Gods. The Gods petitioned Lord Vishnu. Lord Vishnu in the disguise of a beautiful Youth (*i.e.* Daityasudana) won two sisters of Lavanasura. With their assistance, Vishnu discovered his abode, and overcome him in single combat and buried in a pit of his home. The present lake at Lonar is said to be a resort of Lavanasura. It is further said that the water of the lake is the blood and the salt in it is flesh of Lavanasura. Similar legend is also associated with the town of Mehkar. A demon 'Meghaankara' lived at Mehkar. Lord Vishnu, in the disguise of 'Sharangdhara' put the demon to death and the present town Mehkar is the scene of the fight.

In historical times, the district of Buldhana and the adjacent regions were ruled by Mauryas, Satvahanas, Traikutas, Vakatakas, Vishnukundins, Kalachuris and Chalukyas. During the medieval periods, this region was under the regime of the Mohmedan rulers. In the 18th century, it was ruled by Nizam and Shahu, Maratha rulers. The ruling is then followed by the Britishers till 14th August 1947.

10
The People

The district is inhabited by the people of various religions like Hindu, Muslim, Christian, Sikh, Buddhist and Jain. They speak, apart from the state language Marathi, languages or dialects like Banjari, Kachhi, Gujrathi, Hindi, Urdu, Korku, Sindhi and others. They belong to different castes *viz.*, Brahman, Rajput, Wani, Vidur, Gurav, Bairagi, Jogi, Bhat, Kasar, Shimpi, Kunbi, Gawali, Vanjari, Mali, Bari, Lodhi, etc. There are about 15 scheduled castes *viz.*, Balachi, Basor, Bedar, Bhangi, Chambhar, Dohor, Dom, Ganda, Ghasi, Kaikadi, Katia, Khatik, Kori, Mahar and Mang. There are also artisan castes like Sutar, Lohar, Beldar, Kumbhar, Panchal, Ghisadis, Pathrats, Rangari and few others.

The population of Hindus is comparatively larger. They worship gods and goddesses as in the other parts of the state. It is also the case of rituals, rites and ceremonies. The people of other religions also have their own rituals, rites, ceremonies and customs. Hindus celebrate festivals like Gudhipadva, Ramnavami, Hanuman Jayanti, Maha Ekadashi, Gokul Ashtami, Pola, Ganesh Chaturthi, Dasara, Diwali and Holi, etc.

The structure and composition of abodes is also similar to other districts of the state of Maharashtra. The majority of houses are built with burnt bricks, dressed stones, teakwood beams and poles, bamboo ceiling and flat roofs. The building structure is based upon stone foundation, supported by teakwood poles and thick sidewalls. Clay walls are also common. In general, the houses are 'dhaba' type. In modern period, the houses in their structure and composition have largely changed. Some pertinent information of the tehsils under study is provoided in the Table 2.

Table 2: Miscellaneous Information of Tehsils under Investigation

Sl.No.	*Descriptions*	*Figures*
(i) Tehsil: Malkapur		
1.	Area in sq.km.	462
2.	Population	158186
3.	Land Cultivated	42355 hectares
4.	Land Uncultivated	350.32 hectares
5.	Govt. Barren Land	392.52 hectares
6.	Other Barren Land	308.82 hectares
7.	Forest Area	390 hectares
8.	Grazing Land	268 hectares
9.	Irrigated Land	388 hectares
10.	Main Crops	Rice, Jawar, Wheat, Sugarcane, Cotton, few other crops
11.	Cattle	35065
12.	Places of Pilgrims	(03) Malkapur, Kundkhard, Harsoda
13.	Family Planning Centers	03
(ii) Tehsil: Nandura		
1.	Area in sq.km.	387
2.	Population	155454
3.	Land Cultivated	49208 hectares
4.	Land Uncultivated	357.78 hectares
5.	Govt. Barren Land	2141.13 hectares
6.	Other Barren Land	169.05 hectares
7.	Forest Area	162.86 hectares

Contd...

Table 2–Contd...

Sl.No.	*Descriptions*	*Figures*
8.	Grazing Land	269.69 hectares
9.	Irrigated Land	916 hectares
10.	Main Crops	Rice, Jawar, Wheat, Groundnut, Sugarcane, Cotton, Maize, Soyabean and few other crops.
11.	Cattle	49334
12.	Places of Pilgrims	(03) Nandura, Nimbola, Vadner
13.	Family Planning Centers	04
(iii) Tehsil: Motala		
1.	Area in sq.km.	28656
2.	Population	143743
3.	Land Cultivated	57476.89 hectares
4.	Land Uncultivated	126.46 hectares
5.	Govt. Barren Land	4246.05 hectares
6.	Other Barren Land	419.8 hectares
7.	Forest Area	3006 hectares
8.	Grazing Land	Nil
9.	Irrigated Land	Nil
10.	Main Crops	Rice, Jawar, Wheat, Groundnut, Sugarcane, Cotton, Maize, and few other crops
11.	Cattle	Nil
12.	Places of Pilgrims	05 Dhamangaonbadhe, Kothadi, Rohonkhed, Kurhdabazar, Pimpri
13.	Family Planning Centers	04
(iv) Tehsil: Khamgaon		
1.	Area in sq.km.	871.
2.	Population	229295
3.	Land Cultivated	81052.75 hectares
4.	Land Uncultivated	839.81 hectares
5.	Govt. Barren Land	4924 hectares
6.	Other Barren Land	11251 hectares

Contd...

Table 2–Contd...

Sl.No.	Descriptions	Figures
7.	Forest Area	17857.44 hectares
8.	Grazing Land	Nil
9.	Irrigated Land	898 hectares
10.	Main Crops	Rice, Jawar, Groundnut, Sugarcane, Cotton, and few other crops
11.	Cattle	108789
12.	Places of Pilgrims	Nil
13.	Family Planning Centers	07
(v) Tehsil: Shegaon		
1.	Area in sq.km.	505.48
2.	Population	133919
3.	Land Cultivated	45612.63 ht.
4.	Land Uncultivated	459-27 ht.
5.	Govt. Barren Land	761.75 ht.
6.	Other Barren Land	459.27 ht.
7.	Forest Area	499.85 ht.
8.	Grazing Land	62.23 ht.
9.	Irrigated Land	368 ht.
10.	Main Crops	Rice, Jawar, Wheat, Pea nut, Cotton, Sugarcane and other
11.	Cattle	32,160
12.	Places of Pilgrims	Shegaon, Gaygaon, Machhindrakhed, Shri Kshetra Nagzari
13.	Family Planning Centers	05
(vi) Tehsil: Sangrampur		
1.	Area in sq.km.	502.33
2.	Population	120,856
3.	Land Cultivated	435,93 ht.
4.	Land Uncultivated	939-43 ht.
5.	Govt. Barren Land	400-70 ht.
6.	Other Barren Land	99-20 ht.

Contd...

Table 2–Contd...

Sl.No.	Descriptions	Figures
7.	Forest Area	1411 ht.
8.	Grazing Land	1405-38 ht.
9.	Irrigated Land	698 ht.
10.	Main Crops	Rice, Jawar, Wheat, Groundnut, Tur, Banana, Cotton, Sugarcane etc.
11.	Cattle	55036
12.	Places of Pilgrims	Paturda, Warwat-bakal, Sonala, Kankhed
13.	Family Planning Centers	03
(vii) Tehsil: Jalgaon (Jamod)		
1.	Area in sq.km.	589.48
2.	Population	136776
3.	Land Cultivated	41069.59 ht.
4.	Land Uncultivated	168.69 ht.
5.	Govt. Barren Land	4225.92 ht.
6.	Other Barren Land	168.69 ht.
7.	Forest Area	12462.06 ht.
8.	Grazing Land	1021.79 ht.
9.	Irrigated Land	930 ht.
10.	Main Crops	Cotton, Tur, Jawar, Groundnut etc.
11.	Cattle	–
12.	Places of Pilgrims	Dhanora, Palashi-supo, Ellora
13.	Family Planning Centers	–
(viii) Tehsil: Buldhana		
1.	Area in sq.km.	788
2.	Population	2,47,538
3.	Land Cultivated	57960 hectares
4.	Land Uncultivated	3012.52 hectares
5.	Govt. Barren Land	28 hectares
6.	Other Barren Land	58 hectares

Contd...

Table 2–Contd...

Sl.No.	*Descriptions*	*Figures*
7.	Forest Area	11717 hectares
8.	Grazing Land	57.49 hectares
9.	Irrigated Land	10462.2 hectares
10.	Main Crops	Cotton, Groundnut, Soyabean, Jawar, Wheat, Rice, Corn etc.
11.	Cattle	94559
12.	Places of Pilgrims	05 Buldhana, Rajur, Kolwad, Sailani, Girida
13.	Family Planning Centers	Nil
(ix) Tehsil: Chikhali		
1.	Area in sq.km.	797 .
2.	Population	247101
3.	Land Cultivated	112645 hectares
4.	Land Uncultivated	-
5.	Govt. Barren Land	1145 hectares
6.	Other Barren Land	17945 hectares
7.	Forest Area	11091 hectares
8.	Grazing Land	36 hectares
9.	Irrigated Land	-
10.	Main Crops	Cotton, Groundnut, Soyabean, Jawar, Sugarcane, Wheat, Rice, Corn, Kardai etc.
11.	Cattle	90389
12.	Places of Pilgrims	05 Chikhali, Jambhora, Rohda, Kolhara, Ambadapur
13.	Family Planning Centers	Nil
(x) Tehsil: Mehkar		
1.	Area in sq.km.	1131
2.	Population	229621
3.	Land Cultivated	86800 hectares
4.	Land Uncultivated	290.30 hectares
5.	Govt. Barren Land	1221.07 hectares
6.	Other Barren Land	527 hectares

Contd...

Table 2–Contd...

Sl.No.	Descriptions	Figures
7.	Forest Area	12541.62 hectares
8.	Grazing Land	-
9.	Irrigated Land	-
10.	Main Crops	Cotton, Groundnut, Soyabean, Jawar, Peas,Sunflower, Sugarcane, Wheat, Rice, Corn, Kardai etc.
11.	Cattle	91209
12.	Places of Pilgrims	03 Janefal, Gomeshwar, Kalemshwar
13.	Family Planning Centers	-
(xi) Tehsil: Lonar		
1.	Area in sq.km.	67226.8
2.	Population	106628
3.	Land Cultivated	37641 ht.
4.	Land Uncultivated	37141 ht.
5.	Govt. Barren Land	3582.53 ht.
6.	Other Barren Land	3275 ht.
7.	Forest Area	3576.57 ht.
8.	Grazing Land	271.26 ht.
9.	Irrigated Land	1881.24 ht.
10.	Main Crops	Rice, Jawar, Wheat, Sugarcane, Cotton, Groundnut, Tur, Soybean and few other crops
11.	Cattle	93000
12.	Places of Pilgrims	(01) Lonar lake
13.	Family Planning Centers	06
(xii) Tehsil: Sindkhed Raja		
1	Area in sq.km.	78168.64
2	Population	1361152
3	Land Cultivated	37414.02 ht.
4	Land Uncultivated	1341.2 ht.
5	Govt. Barren Land	3899.89 ht.

Contd...

Table 2–Contd...

Sl.No.	Descriptions	Figures
6.	Other Barren Land	32.71 ht.
7.	Forest Area	3899.89 ht.
8.	Grazing Land	3771.42 ht.
9.	Irrigated Land	1971.39 hectares
10.	Main Crops	Rice, Jawar, Wheat, Groundnut, Sugarcane, Cotton and few other crops
11.	Cattle	97371
12.	Places of Pilgrims	(01) Jijau Palace
13.	Family Planning Centers	04
(xiii) Tehsil: Dewulgaon Raja		
1.	Area in sq.km.	69276.8
2.	Population	106126
3.	Land Cultivated	38644.03 ht.
4.	Land Uncultivated	3176 ht.
5.	Govt. Barren Land	3520.57 ht.
6.	Other Barren Land	265.26 ht.
7.	Forest Area	3520.57 ht.
8.	Grazing Land	265.26
9.	Irrigated Land	38644.03
10.	Main Crops	Rice, Jawar, Wheat, Groundnut, Sugarcane, Cotton, Soybean and few other crops
11.	Cattle	10,000
12.	Places of Pilgrims	(01) Balaji Mandir
13.	Family Planning Centers	08

11

Methodology Adapted

Since October 2006, intensive field tours were conducted covering all parts of the tehsils undertaken for this investigation in all seasons. Every attempt has been made to gather information on plant species or their parts/products found useful for local people with reference to medicine, food, fibre, houses, huts, tools and implements useful for agricultural works, etc. During these surveys, methods as suggested by Jain (1987, 1989) have been followed.

The local medicine-men and women, elder persons, farmers, and farm labourers, heads of some communities, etc. were interviewed to collect first-hand information. At some instances, actual applications were also observed and recorded. Information with respect to plant species or their parts used, preparation of recipes, doses, method of administeration, type of diseases treated, etc. were critically and patiently inquired. Plants used for miscellaneous purposes were mostly personally observed and documented. A special care was taken to note local plant names in the area studied.

The plant species are deciphered using floras by Cooke (1958), Naik (1998), Ugemuge (1986), Kamble and Pradhan (1988), Dhore (2002), Kshirsagar and Patil (2008), Patil (2003), Patel (1968), Lakshinarsimhan and Sharma (1991), Singh and Karthikeyan (2000, 2001), etc.

Routine method of herbarium preparation as suggested by Jain and Rao (1977) has been followed. They are given collection numbers and deposited in the Herbarium, P.G. Department of Botany, L.K. Dr. P.R. Ghogrey Science College, Dhule (M.S.).

Ethnobotanical information gathered from area under investigation is presented under the headings *viz.*, Medicine, ethnoveterinary medicine, and miscellaneous uses. The plant species have been arranged alphabetically. Their botanical names, family, local names and photograph numbers, if any, are given at the beginning of the data so accrued. Few photographs, tables and maps are also provided to authenticate the data. The area under investigation was divided, for convenience, into four groups of tehsils is as given under respective Sections I to IV in the ethnobotanical enumeration.

12

Earlier Botanisation in Vidarbha

Buldhana is one of the districts of Vidarbha region of Maharashtra. No studies, to date, paid attention to Buldhana district as an exclusive of topic of research in Ethnobotany. It, therefore, appears pertinent to consider botanisation in this part of the state in the perspective of traditional plantlore. Buldhana and other districts of Vidarbha were once a part of Berar Forest circles of Central Province. The region of Vidarbha has been studied from floristic point of view in pre- and post-independence period (*cf.* Brandis, 1874; Witt, 1911, 1916; Patel, 1968; Graham,1911; Mirashi, 1954, 1960a, 1960b; Mirashi and Nafade, 1961, 1963; Mirashi and Paradkar, 1962; Deshpande and Deshpande, 1963; Dhore 1971, 2000; Donde, 1966; Balapure, 1966, 1971; Malhotra and Moorty, 1971; Purekar, 1985; Kamble and Pradhan, 1988; Ugemuge, 1986; Diwakar and Sharma (2000) etc.). All these attempts tried to reveal the plant wealth of this region.

Ethnobotanical studies are initiated especially in the state of Maharashtra by late Dr. V.D. Vartak. The Vidarbha region has received some attention on this line of research (*cf.* Badhe and Sharma, 1981. Badhe and Pandey, 1988; Badhe, Sharma and Pandey, 1997; Diwakar and Sharma, 2000; Bhogaonkar and Devarkar, 2001a, b; Bhogaokar and Devarkar,2002. Bhogaonkar and Kadam, 2006; Chute and Tiwari, 1999; Das and Misra, 1988; Deshmukh and Dhore, 1994; Chaudhari and Hutke, 2002; Deshmukh, Muratkar and Rothe, 1999; Giri, 1994; Malhotra and Moorthy, 1973; Kamble and Pradhan, 1982.

It appears from the above resume that Buldhana district although has received some attention from floristic point of view, but has remained practically untapped ethnobotanically. This indicated a necessity to study this district as an exclusive topic of research in ethnobotany.

PART – II
ETHNOBOTANICAL ENUMERATION

SECTION–I
(Malkapur, Nandura, Motala & Khamgaon Tehsils)

1. *Abelmoschus manihot* (L.) Medik. (Malvaceae)
Jangali – bhendi:

Ethnoveterinary Medicine

(*i*) Fruits (20-25) are fed mixed with fodder to the pet animals suffering from dysentery. This is followed for 3-4 days.

Miscellaneous Uses

(*i*) Dried plants are used for partition wall of huts.

(*ii*) Young shoots and leaves are used as fodder for cattle.

(*iii*) Stems are used as fuel.

2. *Abrus precatorius* L. (Fabaceae) Gunj

Medicine

(*i*) Seeds are powdered. This powder is mixed with 50gm. powder of seeds of *Corchous depressus* L., 50gm fruit powder of *Emblica officinalis* Gaertn. and about 250 gm candy. A spoonful of it is consumed with milk twice a day to treat spermatorrhoea till cure.

Miscellaneous Uses

(*i*) Goldsmith use seeds for weighing gold (However, this is rarely observed now–a-days).

3. *Abutilon indicum* L. (Malvaceae) Dabba, Atti

Medicine

(*i*) Dried roots are powdered. This powder is homogenized with oil and applied on joints to treat joint-ache.

Miscellaneous Uses

(*i*) Dried plants are used in making of partition walls huts.

(*ii*) Dried stems are used as fuel.

4. *Acacia chundra* Willd. (Mimosaceae) Khair

Miscellaneous Uses

(*i*) Heart wood is sold/used for catechu production.

(*ii*) Wood is especially used for tassel which is used for preparing/cleaning food grains.

(*iii*) Wood is used for making ladder.

5. *Acacia leucophloea* (Roxb.) Willd. (Mimosaceae) Hiwwar

Medicine

(*i*) Fruit paste is prepared in cow urine. It is applied on injury till cure.

6. *Acacia nilotica* (L.) Willd. (Mimosaceae) Babhul, Teli babhul

Medicine

(*i*) Dried leaves are crushed and powdered. It is mixed with some alum. This powder is used to clean tooth and tooth-ache and also to remove foul smell of mouth.

Miscellaneous Uses

(*i*) Wood is used as fuel.

(*ii*) Wood is used for making ploughs and handles of agricultural tools.

(*iii*) Young branches are used for fencing, farmyards and crop fields.

(*iv*) Fruits constitute good fodder for sheeps and goats.

7. *Acanthospermum hispidum* DC. (Asteraceae) Bokharu, Bokhre

Medicine

(*i*) Leaves are burnt and the ash, about two spoonful mixed with coconut oil, is applied on wound till cure.

Ethnoveterinary Medicine

(*i*) Leaf ash is homogenized with coconut oil and applied around horns of pet animals when the animals suffer from itching or worms. These diseases are indicated by the animals while rubbing head/horns against trunks of trees or walls.

8. *Achyranthes aspera* L. (Amaranthaceae) Kharata:

Medicine

(*i*) Fruits of this plant and feathers of peacock are cut into pieces. These are smoked by the sufferer of scorpion– sting.

9. *Aegle marmelos* (L.) Corr. (Rutaceae) Bel

Medicine

(*i*) Fruit pulp is about 20 gm trice a day is consumed to check dysentery. It is beneficial if consumed for 2-3 days.

Miscellaneous Uses

(*i*) Leaves are offered to Lord Shankar by Hindu communities.

10. *Aerva lanata* (L.) Juss. (Amaranthaceae) Safed– Fuli, Pandhari-Fuli

Medicine

(*i*) Extract of leaves, about spoonful twice a day for three days, is orally administered to treat malaria and typhoid.

Ethnoveterinary Medicine

(*i*) Extract of leaves, about one litre, administered orally to pet cows, buffalos and bullocks suffering from fever.

Miscellaneous Uses

(*i*) It is used as fodder for goats.

11. *Agave americana* L. (Agavaceae) Ketki

Medicine

(*i*) Extract of leaves, about spoonful, is mixed with lemon juice. It is taken orally against stomach gases. It is also useful for liver disorders.

12. *Ailanthus excelsa* Roxb. (Simaroubaceae) Mahruph

Medicine

(*i*) Extract of leaves is applied on the stomach to control dysentery in case of children. It is also beneficial as wormicide if given internally in small doses.

Miscellaneous Uses

(*i*) Wood is used in a fuel.

(*ii*) Wood is employed in hut construction.

13. *Albizia lebbeck* (L.) Willd.(Mimosaceae) Shirish

Miscellaneous Uses

(*i*) The leaves are used as fodder for cattle.

14. *Allium cepa* L. (Liliaceae) Kanda

Medicine

(*i*) Bulb is crushed in whey. It is taken orally at early morning for about 15 days to treat piles.

(*ii*) Bulb is directly consumed along with bread prepared from flour of Javari (*Sorghum bicolor* (L.) Moench) to treat scorpion sting.

15. *Allium sativum* L. (Liliaceae) Lasun

Medicine

(*i*) Few bulblets are tied on the neck at night. It is practiced for three days. It is beneficial to check cold.

(*ii*) Bulblets are crushed and the paste is applied on head at night to treat head-ache.

16. *Aloe vera* L. (Liliaceae) Korphad

Medicine

(*i*) Latex is tapped from leaves. About two-three spoons of it consumed at morning for few days against diabetes.

(*ii*) Latex tapped from leaves is mixed with turmeric powder. It is applied on injuries till cure.

17. *Alternanthera sessilis* DC. (Amaranthaceae) Tondalya

Miscellaneous Uses

(*i*) Plants are useful as fodder for cattle.

18. *Alternanthera tenella* Colla (Amaranthaceae) Pandhari-fuli

Medicine

(*i*) Extract of leaves, about 2-3 drops, are applied on the injuries or cuts. This is followed till cure.

Miscellaneous Uses

(*i*) The plant constitutes a good fodder for pet animals.

19. *Amaranthus hybridus* L. (Amaranthaceae) Rajgire, Rajgira

Medicine

(*i*) Roots are crushed and pellets are prepared. Two pellets are consumed for seven days as a remedy against acidity.

(*ii*) Leaves are crushed. This paste is applied on wound till cure.

19. *Annona reticulata* L. (Annonaceae) Ramphal

Medicine

(*i*) Extract of leaves, about spoonful, is taken orally with cow milk. It is drunk against complaints of liver.

20. *Annona squamosa* L. (Annonaceae) Sitaphal

Medicine

(*i*) Dried fruit is powdered. It is homogenized in honey. Two spoons of it are consumed thrice a day for 24 days to treat asthma.

(*ii*) Leaves are crushed and tied on the injuries. It is wormicidal in action and is applied till cure.

Ethnoveterinary Medicine

(*i*) Leaves are powdered. It is applied on injuries for deworming.

Miscellaneous Uses

(*i*) Wood is used as a fuel.

21. *Anogeissus latifolia* (Roxb. ex DC.) Wall. (Combrataceae) Dhawda

Medicine

(*i*) Stem bark is powdered. About spoonful is consumed along with single flower of *Catharanthus roseus* (L.) G. Don. This is followed for about 15 days to control diabetes.

22. *Apluda mutica* L.(Poaceae) Shippi-Gawat

Miscellaneous Uses

(*i*) Plants are used as a fodder.

(*ii*) Dried plants are used for making partition wall of huts.

23. *Argemone mexicana* L. (Papaveraceae) Satyanashi; Vilayachi

Medicine

(*i*) Seeds are heated and then powdered. The paste is prepared in water. It is applied on injuries caused due to piles.

(*ii*) Roots are burnt; fumes are passed on wound caused due to piles. It is practiced once daily for 3-4 days.

24. *Aristolochia bracteolata* Lamk. (Aristolochiaceae) Gindiyan

Medicine

(*i*) Extract of leaves is mixed in coconut oil. It is prepared with camphor. Paste is applied on injury caused due to piles.

(*ii*) Leaf extract is prepared in water. About 2-3 drops are dropped into ears to check pus formation.

(*iii*) Leaf juice is taken orally about spoonful twice a day for seven days to treat migraine.

Ethnoveterinary Medicine

(*i*) Extract of leaves is applied to deworm wounds of cows, bullocks and buffaloes.

25. *Asparagus racemosa* Willd. (Liliaceae) Shatawari

Ethnoveterinary Medicine

(*i*) Root of this plant species and also of *Withania somnifera* (L.) Dunal are dried and powdered. About 250 gm of this powder is dissolved in one litre of water. It is administered orally to the milching buffaloes to increase milk flow.

26. *Asphodelus tenuifolius* Cav. (Liliaceae) Zar-kanda

Medicine

(*i*) Leaves are heated and oil is applied on hot leaves. These are wrapped around the suppurating tumor for 3-4 days.

(*ii*) Decoction of leaves is applied on fingers to treat whitlow.

(*iii*) Extract of leaves is applied on burns. It is followed till cure.

27. *Azadirachta indica* A. Juss. (Meliaceae) Neem

Medicine

(*i*) Extract of stem bark, about a cup, is drunk at morning for 15 days in case of heart problem.

(*ii*) Leaves are boiled in water. This decoction is used while bathing as a remedy against itching.

(*iii*) Extract of leaves is given orally to the patient suffering from fever. Each dose contains spoonful of extract daily to cure.

(*iv*) Seeds after removal of seed coats are crushed and mixed with ammonia salt (in 4:1 ratio). Pellets are prepared. A pellet is advised twice a day against piles.

Miscellaneous Uses

(*i*) Wood is used in house construction.

(*ii*) Wood is used in making agricultural tools.

28. *Balanites aegyptiaca* (L.) Del. (Balanitaceae) Hengan-Beth

Medicine

(*i*) Pulp of fruit is consumed along with single leaf of betel (*Piper betle* L.) It is consumed thrice a day for few days against cough.

Ethnoveterinary Medicine

(*i*) Pulp of fruit is crushed in water. About 2-3 drops are applied on eye injuries till cure.

Miscellaneous Uses

(*i*) Ripe fruit is crushed on water. This mixture is used as a detergent.

29. *Bambusa arundinacea* Wild. (Poaceae) Bamboo, Tokar

Miscellaneous Uses

(*i*) Leaves are used in making roof and partition wall of huts.

(*ii*) Stem is employed for making baskets and bullock-carts.

(*iii*) Bamboo culms are employed for making ladder.

30. *Basella alba* L. (Basellaceae) Vavad-hing, Vavading

Medicine

(*i*) Two to three seeds are boiled in milk. About spoonful of it is given orally to children suffering from stomach-ache.

(*ii*) Seeds are powdered and mixed with jagerry 1:5 ratio consumed for three to four days for deworming.

Ethnoveterinary Medicine

(*i*) Seeds are powdered and mixed with jaggery (in 1:5 ratio) and given to pet animals for three to four days against worms. About 250 gm of it is fed per day.

31. *Biophytum sensitivum* (L.) DC. (Oxalidaceae) Sharmily, Lajalu

Medicine

(*i*) A spoonful of root extract is taken orally twice a day to treat diphtheria. It is advised for a week or so.

32. *Brassica juncea* L. (Brassicaceae) Gawat

Medicine

(*i*) Flowers are crushed in water with few grains of wheat. The paste is applied on suppurating tumours.

33. *Butea monosperma* (Lamk.) Taub. (Fabaceae) Palas

Medicine

(*i*) Green fruits are dried and powdered along with a seed of *Tectona grandis* L. This paste is applied on abdominal region as a remedy against flatulence. It is advised for 2-3 days.

Miscellaneous Uses

(*i*) Leaves are used for preparing dining plates and bowls.

(*ii*) Leaves are used as fodder for pet animals.

(*iii*) Wood and leaves are used for hut construction.

34. *Caesalpinia bonduc* (L.) Roxb. (Caesalpiniaceae) Sagargoti

Medicine

(*i*) Seeds are rubbed in water. This extract is administered to the children for deworming. It is advised till cure.

35. *Cajanus cajan* (L.) Millsp. (Fabacea) Tur

Miscellaneous Uses

(*i*) Stems are used for sweeping ground and farmyard.

(*ii*) Young shoots and foliage constitute good fodder for domestic animals.

36. *Calotropis gigantea* (L.) R.Br. (Asclepiadaceae) Rui

Medicine

(*i*) Latex is mixed with jaggery. Pellets are prepared. Single pellet is consumed twice a day for two days to treat stomach ache.

Ethnoveterinary Medicine

(*i*) Latex is homogenized with red lead (5gm) and groundnut oil. It is applied below the tail of a pet animal to treat dullness of eye-vision.

(*ii*) Leaves (5-6) are fed to treat animal suffering from flatulence.

37. *Calotropis procera* (A.t.) R. Br. (Asclepiadaceae) Rui

Medicine

(*i*) Root bark is consumed with betel leaf (*Piper betle* L.) treat heart-ache.

(*ii*) Latex is applied on injury caused due to thorns.

Ethnoveterinary Medicine

(*i*) Latex is applied on leaves. The leaves are heated and wrapped around the fractured organ/bone.

38. *Canna indica* L. (Cannaceae) Kardal

Medicine

(*i*) Leaves are boiled in four cups of water. It is boiled till the decoction is reduced to one cup. One spoonful of it is administered twice a day for 8-10 days against urinary stone.

39. *Capparis decidua* (Forsk.) Edgw. (Capparadaceae)

Miscellaneous Uses

(*i*) Twigs are used for fencing farmyard.

40. *Capparis zeylanica* L. (Capparidaceae) Fulya–yel

Ethnoveterinary Medicine

(*i*) Flowers and leaves are crushed. The paste is applied on wound to avoid infection from germs.

Miscellaneous Uses

(*i*) It is used in the construction of huts.

(*ii*) It is a used as fodder especially for goats.

41. *Capsicum annuum* L. (Solanceae) Mirchi

(*i*) Ripe dried fruits are powdered. It is applied on injuries caused due to dog–bite.

42. *Cardiospermum halicacabum* L. (Sapindaceae) Kapal-phodi

Miscellaneous Uses

(*i*) The plants are used as fodder for goats.

43. *Carica papaya* L. (Cariaceae) Papai

Medicine

(*i*) Fruit latex (5 to 10 ml) is taken orally for 2-3 days to regulate menstrual cycle.

(*ii*) Ripe fruit is consumed at early morning for seven days to reduce blood-pressure.

44. *Cassia obtusifolia* L. (Caesalpinaceae) Tarota

Medicine

(*i*) Leaves are crushed. The paste is tied on joints to cure joint-ache till cure.

(*ii*) Seeds are powdered and bread is prepared. It is consumed for seven days to treat joint-ache.

45. *Cassia occidentalis* L.(Caesalpiniaceae) Dev-tarota

Ethnoveterinary Medicine

(*i*) Extract of leaves is applied on wounds to deworm till cure.

46. *Catharanthus roseus* (L.) G. Don (Apocynaceae) Sadabahar

Medicine

(*i*) To cure head-ache, the stems are broken and then smelt.

47. *Celosia argentea* L. (Amaranthaceae) Killu, Karadu

Medicine

(*i*) Root extract is prepared in water. Two-three spoons of it are given orally 2-3 times a day to cure sun- stroke.

(*ii*) Root extract is given orally. One spoonful is administered twice a day to cure hepatitis. It is administered till cure.

Ethnoveterinary Medicine

(*i*) Roots are crushed in coconut oil. It is diluted in water. About one liter of it is administered to treat emergence of placenta after delivery in case of cows.

Miscellaneous Uses

(*i*) It is used as fodder for domestic animals.

48. *Chenopodium murale* L. (Chenopodiaceae) Chil

Miscellaneous Uses

(*i*) Leaves are used as vegetable.

(*ii*) Plants are used as fodder for cattle.

49. *Chlorophytum borivilianum* Santapdu and Fernand. (Liliaceae) Safed-musali

Medicine

(*i*) Roots are crushed. The juice is mixed with whey. About 2-3 spoons is drunk daily to control body heat. It is advised for seven days.

(*ii*) Root is powdered. About one spoonful of it is consumed with milk to treat heart-ache.

50. *Cicer arientinum* L. (Fabaceae) Harbhara

Medicine

(*i*) Seeds are soaked in water and are consumed for 3-4 days to kill intestinal worms.

(*ii*) Extract of leaves, about spoonful, is taken orally twice a day for two days to treat stomach-ache.

Ethnoveterinary Medicine

(*i*) She-buffaloes are fed with its pulse along with jaggery (*Saccharum officinarum* L.) and sago after delivery to expel foetus early.

51. *Cissus quadrangula* L. (Vitaceae) Kandyawel

Medicine

(*i*) Stem is rubbed in water. This paste is applied on finger to cure whitlow till cure.

(*ii*) Stem is crushed and the extract obtained about 4-5 ml is administered for normal delivery.

Ethnoveterinary Medicine

(*i*) Stem is crushed and applied on fractured bone till cure.

52. *Citrullus colocynthis* (L.) Schrad. (Cucurbitaceae) Kadu-indrawan

Medicine

(*i*) Extract of leaves, about spoonful, is given orally twice or thrice a day for three days to cure stomach-ache.

53. *Citrullus lanatus* (Thunb.) Matsumura and Nakai. (Cucurbitaceae) Tarbuj

Medicine

(*i*) Seeds are powdered. It is applied on wound caused due to spider bite or scorpion bite.

54. *Citrus aurantifolia* (Christm.) Sw. (Rutaceae) Limbu

Medicine

(*i*) Fruit juice is diluted in water. About 200 ml is administered twice a day to cure flatulence.

(*ii*) Leaf extract is prepared in water. It is applied on body to reduce itching.

(*iii*) Fruit juice is applied on hair at night for 3-4 days to control dandruff.

Ethnoveterinary Medicine

(*i*) Fruit juice is given orally to a calf suffering from flatulence. About 50 ml of juice is diluted in half litre and administered once a day for 2-3 days.

55. *Citrus sinensis* L. (Rutaceae) Mosambi

Medicine

(*i*) Fruit juice is taken orally by a woman, about a cup, for seven days to increase flow of milk.

56. *Cleome gynandra* L. (Capparidaceae) Pandhari-tilwan

Medicine

(*i*) Leaf extract, few drops, are poured into nose against head-ache.

(*ii*) Extract of leaves, about 2-3 drops, are dropped into ears for 2-3 days against molar-ache.

57. *Clerodendrum multiflorum* (Burm *f.*) O. Ktze. (Verbanaceae) Arni

Medicine

(*i*) Leaves are crushed. The juice obtained, about spoonful, is administered to children suffering from intestinal worms.

Miscellaneous Uses

(*i*) Dried plants are used in making of partition walls and roof of huts.

(*ii*) Young branches are used as fodder particularly for goats.

(*iii*) Dried branches are used as fuel.

58. *Clitoria ternatea* L. (Fabaceae) Gokarni

Miscellaneous Uses

(*i*) The plants are used as fodder for goats.

59. *Coccinia grandis* Voigt. (Cuccurbitaceae) Tongla, Tondla

Miscellaneous Uses

(*i*) Plants are fed to goats.

60. *Cocos nucifera* L. (Arecaceae) Naral

Medicine

(*i*) Pulp of dried kernel is powdered. It is consumed to cure mouth ulcer.

61. *Colocasia esculenta* (L.) Schott. (Araceae) Alu-pan, Pothi

Medicine

(*i*) Extract of leaves, about two spoons, drunk twice a day to treat urine-stone for 7-8 days.

62. *Commelina forskalaei* Vahl (Commelinnaceae) Keni

Ethnoveterinary Medicine

(*i*) Plants are crushed. Turmeric powder is added in it and thoroughly stirred. It administered to goats to treat cough.

63. *Commicarpus chinensis* (L.) Heimerl. (Nyctaginaceae) Nil

Miscellaneous Uses

(*i*) Plants are fed to goats.

64. *Convolvulus arvensis* L. (Convolvulaceae) Sadhiwasan

Ethnoveterinary Medicine

(*i*) Extract of leaves is applied on injury to avoid infection from worms.

Miscellaneous Uses

(*i*) It is used to tie bundles of fodder.

(*ii*) It is used as a fodder as well.

65. *Corchorus depressus* L. (Tiliaceae) Harankhuri

Medicine

(*i*) Entire plant is crushed with whey. About one spoonful of it is taken orally twice a day to treat burnings of eyes and stomach.

(*ii*) Extract of leaves, 2-3 spoons, is taken orally with whey twice a day for 15-20 days to treat spermatorrhea. It is followed till cure.

66. *Cordia gharaf* (Forsk.) Ehrenb. and Asch. (Boraginaceae) Gondan, Bhokar

Medicine

(*i*) Extract of leaves, about spoonful, given orally twice a day for two days against amoebic dysentery.

Miscellaneous Uses

(*i*) Wood is used for making different utensils.

(*ii*) Fruits are edible.

(*iii*) Leaves are used as feed for goats.

(*iv*) Wood is used as small timber e.g. for agricultural implements and hut constructions.

67. *Coriandum sativum* L. (Apiaceae) Kothambir, Kothmer

Ethnoveterinary Medicine

(*i*) Extract of about half kg entire plants are advised twice a day to cure 'Lal-khurguti' i.e. FMD (foot and mouth disease).

(*ii*) Seeds are crushed and the paste is applied on tongue to cure 'Lal-khurguti' i.e. FMD (foot and mouth disease).

68. *Crotalaria filipes* Bth. (Fabaceae) Fatfati

Miscellaneous Uses

(*i*) Plants are useful as fodder for cattle.

69. *Ctenolepis garcinii* (Burm. *f.*) Naud. (Cucurbitaceae)

Miscellaneous Uses

(*i*) Plants are useful as fodder for goats.

70. *Cucumis calossus* (Rottl.) Cogn. (Cucurbitaceae) Dendolya

Medicine

(*i*) Fruits are crushed with cow-urine and paste is prepared. It is applied to treat mumps.

(*ii*) Leaves are crushed and the paste obtained is applied on wound. It is followed till cure.

Ethnoveterinary Medicine

(*i*) Extract of leaves is applied on wounds till cure.

Miscellaneous Uses

(*i*) It is used as fodder to cattle.

71. *Cucurbita moschata* (Duch. ex Lamk.) Poir. (Cucurbitaceae) Gangaphal

Ethnoveterinary Medicine

(*i*) Fruit is cut into pieces and boiled in water. This water, about 2-3 litres, is administered daily to milching animals for about 10-15 days to promote milk.

72. *Cullen corylfolia* (L.) Medic. (Fabaceae) Fulore

Ethnoveterinary Medicine

(*i*) Extract of leaves is applied on wound for deworming till cure.

Miscellaneous Uses

(*i*) Dried plants are used in making roof and partition walls of huts.

(*ii*) Green plants are used as fodder for cattle.

73. *Curcuma amada* Roxb. (Zingiberaceae) Ambehalad

Medicine

(*i*) Tubers are powdered. This powder and sugar is mixed in flour of wheat. Sweet balls are prepared. One sweet ball each day at morning is consumed to treat rheumatism.

(*ii*) Tuber is powdered. It is mixed in coconut oil and applied on inflamed body part till cure.

74. *Curcuma longa* L. (Zingiberaceae) Halad

Medicine

(*i*) Rhizomes are powdered. It is mixed with lime Two or three spoonful given orally twice a day to control dysentery.

Ethnoveterinary Medicine

(*i*) Stem powder is applied on injury to avoid infection.

75. *Cuscuta chinensis* Lamk. (Cuscutaceae) Amarwel

Medicine

(*i*) Decoction of stem, about spoonful, is taken orally twice a day to treat fever till cure.

76. *Cyamopsis tetragonoloba* (L.) Taub. (Fabaceae) Gawar

Medicine

(*i*) Seeds are boiled in water. Extract of these seeds are, about spoonful, is taken orally to cure diabetes. It is advised for a fortnight.

(*ii*) Mature seeds are powdered. These are consumed, about 10 gm. per day, for 2-3 days to avoid bleeding from wounds.

77. ***Dactyloctenium aegptium*** **(L.) P. Beauv. (Poaceae) Gawat**

Miscellaneous Uses

(*i*) Plants are used as fodder for pet animals.

78. ***Dalbergia latifolia*** **Roxb. (Fabaceae) Sisam**

Miscellaneous Uses

(*i*) Wood is used for making door and window frames.

(*ii*) It is employed as pillars of houses.

(*iii*) It is also employed in making tables, chairs and cots.

(*iv*) Leaves are used as fodder for goats.

79. ***Datura innoxia*** **Mill. (Solanaceae) Dhotara**

Ethnoveterinary Medicine

(*i*) Fruit walls are heated and powdered after cooling. It is mixed in flour and fed to cows, bullocks and buffaloes suffering from dysentery. It is followed for few days.

80. ***Datura metel*** **L. (Solanaceae) Kala-dhotara**

Medicine

(*i*) Leaves are crushed. The leaf juice obtained is applied on throat as a remedy against throat swelling. It is applied till cure.

(*ii*) Leaf paste is applied on tumours and also on mumps.

81. ***Daucus carota*** **L. (Apiaceae) Gajar**

Medicine

(*i*) Decoction of seeds,(5gm)is taken orally for seven days to regulate menstrual cycle.

(*ii*) Extract of root, about a cup, is drunk for fortnight days to treat color blindness.

82. ***Delonix elata*** **(L.) Gamble (Caesalpiniaceae) Sansada**

Miscellaneous Uses

(*i*) Wood is employed as pillars of huts.

(*ii*) It is used as fuel.

(*iii*) Leaves and green fruits are used as fodder especially for goats.

83. *Delonix regia* (Boj. ex Hook.) Raf. (Caesalpiniaceae): Gulmohar

Miscellaneous Uses

(*i*) Wood is used as small timber in preparing e.g. planks find place for domestic utilities.

(*ii*) Wood is used as fuel.

84. *Desmodium velutinum* (Willd.) DC. (Fabaceae)

Miscellaneous Uses

(*i*) Plants are used as fodder for cattle.

85. *Digera muricata* (L.) Mart. (Amaranthaceae) Chhota-kurdu

Miscellaneous Uses

(*i*) Plants are useful as fodder for cattle.

86. *Dinebra retroflexa* (Vahl) Panz. (Poaceae) Ghas-gawat

Miscellaneous Uses

(*i*) It is used as fodder, especially for milching animals.

(*ii*) Dried plants is used for roof of huts.

(*iii*) During 'Haldi' ceremony of marriage in Hindus, plants are kept erect in pandal.

87. *Diplocyclos palmatus* (L.) Jeffery (Cucurbitaceae) Shivalingi

Medicine

(*i*) Extract of seeds (4-5) is prepared in mothers milk spoonful extract is given orally 2-3 times a day to cure flatulence in case of children.

(*ii*) One seed is consumed with a spoonful of animal ghee to cure stomach-ache also for conception. It is advised for seven days.

88. *Eclipta prostrata* (L.) L. (Asteraceae) Maka, Kala-makha

Medicine

(*i*) Leaves are crushed. The paste is applied on wounds caused due to axima.

89. *Ehertia aspera* Roxb. (Boraginaceae) Lokhandi

Ethnoveterinary Medicine

(*i*) Leaves are crushed, fed to milching she-buffaloes to increase milk yield.

Medicine

(*i*) Leaf paste is applied on joints to cure joint-ache.

90. *Emblica officinalis* Gaertn. (Euphorbiaceae) Awala

Medicine

(*i*) Leaves are burnt and powdered. It is mixed with coconut oil and applied on injury caused due to burns.

Miscellaneous Uses

(*i*) Fruit is powdered and used for washing hair and to control dandruff.

91. *Emblica sonchifolia* (L.) DC. (Asteraceae) Mhatara

Miscellaneous Uses

(*i*) Plants are useful as fodder for cattle.

92. *Eucalyptus maculata* Hook. (Myrataceae) Nilgiri

Medicine

(*i*) Leaves and fruits are rubbed in water. The paste is applied on head against head-ache.

Miscellaneous Uses

(*i*) Wood is employed in the construction of huts.

(*ii*) Wood is used for making agricultural equipments, especially ploughs, seed-drills and cultivators.

93. *Euphorbia indica* Lamk. (Euphorbiaceae) Dudhi-gondan

Medicine

(*i*) Latex is mixed with whey. Pellets are prepared. Single pellet is taken twice a day for seven days to treat hepatitis.

Ethnoveterinary Medicine

(*i*) Extract of leaves is mixed with coconut oil (1:1 ratio). It is applied on injury till cure.

94. *Euphorbia millii* Desmoul. (Euphorbiaceae) Katya-thor

Medicine

(*i*) Stem is baked in hot ash. The juice is then extracted from it. About spoonful is drunk for three times a day to cure cough till cure.

Miscellaneous Uses

(*i*) It is used in making of compound.

95. *Euphorbia tirucalli* L. (Euphorbiaceae) Kshar

Medicine

(*i*) Latex is applied on joints to treat joint-ache. It is practiced till cure.

(*ii*) Latex is applied on fingers suffering from whitlo till cure.

Ethnoveterinary Medicine

(*i*) Young branches are crushed and mixed with soil of ant-hill. It is applied on bone fracture and cramps.

96. *Ficus benghalensis* L. (Moraceae) Wad

Medicine

(*i*) Latex is applied on gums to treat tooth-ache till cure.

(*ii*) Latex is consumed with sugar (one spoonful) at morning against impotency for a month.

Ethnoveterinary Medicine

(*i*) Leaves are heated. Oil is applied on them. These leaves are wrapped around bone facture in case of pet animals till cure.

97. *Ficus carica* L. (Moraceae) Angir

Medicine

(*i*) Fruits are consumed daily to increase blood.

(*ii*) Latex is applied on molar to treat molar-ache.

98. *Ficus racemosa* L. (Moraceae) Umber

Medicine

(*i*) Latex is tapped from leaves. It is applied on tooth to treat tooth-ache till cure.

99. *Gliricidia sepium* (Jacq.) Kunth ex Walp.(Fabaceae)

Miscellaneous Uses

(*i*) Entire plants are used for making partition walls of huts.

(*ii*) Seed oil is used in cooking.

100. *Gomphrena globosa* L. (Amaranthaceae) Supari

Medicine

(*i*) Paste is prepared in water of the fruiting head. It applied on head to treat migraine till cure.

101. *Gossypium hirsutum* L. (Malvaceae) Kapus

Miscellaneous Uses

(*i*) Entire plants are used in making roof or partition walls of huts.

(*ii*) Plants are used as fuel.

(*iii*) They are also employed in making baskets useful for agricultural works.

(*iv*) Leaves are useful as fodder.

102. *Grewia tiliaefolia* Vahl (Tiliaceae) Dhaman

Miscellaneous Uses

(*i*) Leaves are used as fodder for goats.

103. *Guazuma ulmifolia* Lamk. (Sterculiaceae) Rudraksha

Miscellaneous Uses

(*i*) Wood is used in making of frames of windows and doors.

104. *Hardwickia binata* Roxb. (Caesalpiniaceae) Anjan

Medicine

(*i*) Leaves are baked and then crushed. The paste is prepared in coconut oil. It is applied on burns or white spots developed on skin after burning.

Miscellaneous Uses

(*i*) Wood is used in making cots and frame of doors and windows.

105. *Helianthus annuus* L. (Asteraceae) Suryphul

Medicine

(*i*) Leaves are crushed. The paste is applied on cramps.

Miscellaneous Uses

(*i*) Dried entire plants are used for partition walls of huts.

106. *Helicteres isora* L. (Sterculiaceae) Mural-sheng

Medicine

(*i*) Fruit paste is prepared in water. It is consumed twice a day for three days to treat stomach-ache. Paste of 4-5 fruits and of one fruit is advised for adults and infants respectively.

107. *Hibiscus cannabinus* L. (Malvaceae) Ambadi

Miscellaneous Uses

(*i*) Fibres are obtained from plants which are used in preparing ropes.

(*ii*) Its fibre is also useful for tying bullocks and carts.

108. *Hibiscus schizopetatus* Hook *f*. (Malvaceae) Jaswand

Medicine

(*i*) Extract of flowers is mixed with sugar. Half cup of it is drunk for two-three days to treat pains in backbone.

(*ii*) Leaves (4-5) are consumed for 15-20 days to cure spermatorrhoea.

109. *Holoptelea integrifolia* (Roxb.) Planch (Ulmaceae) Sanjari

Miscellaneous Uses

(*i*) Seeds are consumed by native people.

(*ii*) Leaves are used as fodder for cattle.

110. *Hordeum vulgare* L. (Poaceae) Javas

Medicine

(*i*) Seeds are crushed and boiled. This paste is applied as the chest to cure cough till cure.

111. *Ipomoea carnea* Jacq. *fistulosa* (Mart.ex Choisy) Austin (Convolvulaceae) Nilwarni

Medicine

(*i*) Leaves are crushed and the paste is applied on cuts till cure.

Ethnoveterinary Medicine

(*i*) Entire plants are crushed. The paste is mixed with soil and applied on fractured bones or cramps.

Miscellaneous Uses

(*i*) Plants are used as fodder for goats.

(*ii*) Entire stems are employed to tie bundles of fodder.

112. *Ipomoea hederifolia* L. (Convolvulaceae) Dudhi-wel

Miscellaneous Uses

(*i*) Plants are useful as fodder for goats.

113. *Ipomoea obscura* (L.) Ker.-Gawl. (Convolvulaceae) Pungali

Miscellaneous Uses

(*i*) Plants are useful as fodder for goats.

114. *Jasminum officinale* L. (Oleaceae) Chameli

Medicine

(*i*) Young leaves are chewed, however the saliva is spat after chewing. This helps cure mouth-ulcer.

115. *Jasminum sambac* (L.) Ait. (Oleaceae) Mogra

Medicine

(*i*) Extract of leaves and flowers is applied as a remedy against scabies and itching. It is practiced till cure.

116. *Jatropha gossypifolia* L. (Euphorbiaceae) Chandrajyoti

Medicine

(*i*) Fruits (2-3) are consumed once a day for two days against dysentery.

117. *Justicia adhatoda* L. (Acanthaceae) Adulsa

Medicine

(*i*) Extract of leaves, about a spoonful, taken orally to cure cough till cure.

(*ii*) Leaves are heated and tied on joints to treat joint-ache.

118. *Kalanchoe pinnata* (Lamk.) Pers. (Crassulaceae) Panphuti

Medicine

(*i*) Extract of leaves, about a spoonful, given orally, twice a day to treat urine-stone. It is advised for about a week.

119. *Kirganelia reticulata* (Poir.) Baill. (Euphorbiaceae) Kamini

Miscellaneous Uses

(*i*) Plants are fed to goats.

120. *Lablab purpurious* (L.) Sweet (Fabaceae) Wal

Medicine

(*i*) Leaves are crushed. The paste is mixed with the paste of leaves of *Lawsonia inermis* L. It is then applied on head for proper growth of hair.

121. *Lagenaria siceraria* (Molina) Standl. (Cucurbitaceae) Dudhi-bhopala

Medicine

(*i*) Extract of leaves, about a cup, is taken orally for few days to treat urinary-stone.

(*ii*) Fruit is cooked and consumed against stomach complains.

122. ***Lagerstroemia parviflora*** **Roxb. (Lythraceae) Bondra**

Medicine

(*i*) Leaves are crushed. The paste is applied on the wound till cure.

123. ***Lantana camara*** **L. (Verbanceae) Kamini**

Medicine

(*i*) Extract of leaves, 1-2 drops, are poured into ears to control ear-ache. It is used till cure.

Miscellaneous Uses

(*i*) Stems are used for partition wall of huts and gates of farmyards.

(*ii*) They are also used as fuel.

124. ***Leucaena latisiliqua*** **(L.) Gill. (Mimosaceae) Subabhul**

Miscellaneous Uses

(*i*) Leaves and green fruits are used as fodder for goats, cows and bullocks.

(*ii*) Wood is used as fuel.

(*iii*) Wood is employed as pillars of huts.

125. ***Leucas cephalotes*** **(Roxb. ex Roth) Spr. (Lamiaceae)**

Miscellaneous Uses

(*i*) Plants are useful as folder for cows and buffaloes.

126. ***Limonia acidissima*** **L. (Rutaceae) Kaut**

Medicine

(*i*) Young leaves (4-5) are chewed twice or thrice a day to treat acidity. The saliva so produce is swallowed.

Miscellaneous Uses

(*i*) Dried branches are fuel.

(*ii*) Wood is used as poles in hut construction.

127. *Lindernia parviflora* (Roxb.) Haines (Scrophulareaceae) Bhui-awala

Ethnoveterinary Medicine

(*i*) Fruits are powdered. The powder is mixed in lemon juice. About 100 ml of it is diluted in one litre water. It is then administered to animals suffering from flatulence daily for 3-4 days.

128. *Luffa acutangula* (L.) Roxb. (Cuccurbitaceae) Dodake

Medicine

(*i*) Leaves are crushed and leaf juice is obtained. It is used to cooked rice. It is then consumed to treat hepatitis for about a week.

129. *Madhuca longifolia* (Koen.) Macbr. (Sapotaceae) Mahu, Moha

Medicine

(*i*) Extract of seeds is given orally twice times a day to cure stomach-ache.

(*ii*) Seeds are consumed along with betel leaf (*Pipper betel* L.) to treat stomach-ache.

Miscellaneous Uses

(*i*) Flowers are soaked in water. These are then used for preparing wine.

130. *Mangifera indica* L. (Anacardiaceae) Amba

Medicine

(*i*) Gum is homogenized with water. It is applied to cure deformations on foot sole.

(*ii*) Latex obtained from unripe fruit is applied after scorpion sting.

Miscellaneous Uses

(*i*) Timber is used in making plants useful in house construction.

(*ii*) Leaves are used in various worships especially by the Hindus.

131. *Manikara zapota* (L.)P. van Royen (Sapotaceae) Chiku

Ethnoveterinary Medicine

(*i*) Leaves and seeds are crushed with soil of ant-hill. The paste is applied on bone facture for 4-5 days.

Miscellaneous Uses

(*i*) Wood is used as fuel.

132. *Maytenus emarginata* (Wild.) Ding Hau (Celastraceae) Henkal

Medicine

(*i*) Young leaves are chewed 2-3 times a day to cure mouth-ulcer. This is continued till cure.

133. *Mentha spicata* L. (Lamiaceae) Pudina

Medicine

(*i*) Extract of leaves, about spoonful, is drunk twice a day to cure hepatitis. It is advised for 7-8 days.

134. *Merrremia emarginata* (Burm.*f*.) Hall. *f*. (Convolvulaceae) Pingli

Medicine

(*i*) Crushed leaves are applied on body against boils.

(*ii*) Leaves are crushed and applied on suppurating tumours till cure. Fresh leaves are used each day.

135. *Mimosa hamata* Willd. (Mimosaceae) Arti

Miscellaneous Uses

(*i*) Leaves are useful as fodder for goats.

136. *Momordica charantia* L. (Cucurbitaceae) Karle

Medicine

(*i*) Extract of leaves, about half cup, is taken orally for seven days to treat diabetes.

137. *Momordica dioica Roxb.* ex Willd. (Cucurbitaceae) Katerle

Ethnoveterinary Medicine

(*i*) Roots are cut into small pieces. These are mixed in fodder and fed to the pet animals. This helps to cure flatulence. It is followed till cure.

138. *Morinda pubescens* J.E. Sm. (Rubiaceae) Aal

Medicine

(*i*) Decoction of leaves, about a cup, is drunk for 14 days to control diabetes.

(*ii*) Fruit is baked. It is kept in mouth near molar to treat molar ache.

139. *Moringa concanensis* Nimmo ex Dalz. and Gibbs. (Moringaceae)

Kadu-shewag

Medicine

(*i*) Flowers are cooked and consumed daily for a month to treat joint-ache.

(*ii*) Roots are crushed. The juice is homogenized in oil. It is applied on joints to treat joint-ache.

Miscellaneous Uses

(*i*) Wood is used as fuel.

(*ii*) Wood is also employed as pillars of huts.

140. *Moringa oleifera* Lamk. (Moringaceae) Sewaga

Medicine

(*i*) Extract of leaves, about spoonful, is taken orally at morning for few days to control diabetes.

(*ii*) Leaves are powdered along with leaves of *Azadiracta indica A. Juss.* and *Dolichandrone falcata* (Wall. ex DC.) Sem. This powder is burnt and the fumes are passed on body part suffering from aching.

141. ***Mucuna pruriens*** **(L.) DC. (Fabaceae) Kach-kuiri**

Medicine

(*i*) Fruits spines are crushed finely and mixed in jaggery. Pellets are prepared. Two pellets are consumed a day for intestinal deworming.

142. ***Murraya koenigii*** **(L.) Spreng. (Rutaceae) Kadi-patta**

Medicine

(*i*) Leaves are washed and fried in ghee. These (8-10) are consumed for few days against body-ache.

143. ***Musa paradisiaca*** **L. (Musaceae)Keli**

Medicine

(*i*) Fruit wall is tied on the suppurating tumour.

(*ii*) Ripe fruit pulp is mixed with milk and seed powder of *Tamarindus indica* L. is consumed to treat impotency.

(*iii*) Fruit is consumed with sulphur to treat itching.

144. ***Nelumbo nucifera*** **Gaertn. (Nymphaeaeae) Kamal**

Medicine

(*i*) Rhizomes paste is applied on wound caused due to piles. It is advised till cure.

(*ii*) Extract of rhizome about spoonful is given orally twice a day for two days against dysentery.

145. ***Nerium indicum*** **Mill. (Apocynaceae) Kanher**

Medicine

(*i*) Flowers and roots are crushed. The paste is prepared in cow-urine. It is applied on injuries caused due to piles.

146. ***Nicotiana tabacum*** **L. (Solanaceae) Tambhakhu**

Medicine

(*i*) Leaves are dried, crushed and powdered. It is homogenized with some lime. It is applied on injury. It acts as wormicide.

147. ***Nyctanthes arbor-tristis*** **L. (Oleaceae) Parijat**

Medicine

(*i*) Stem bark is powdered. Pellets are prepared in ghee. Three pellets a day are advised for 21 days to combat asthma.

148. *Ocimum americanum* L. (Lamiaceae) Ran –tulasi

Medicine

(*i*) Few seeds are kept in milk overnight. This spoonful milk is taken orally for 7 days to cure leucorrhoea.

(*ii*) Decoction of leaves, about two spoonful, is given orally twice a day for three days to treat cough and fever.

149. *Ocimum basilicum* L. (Lamiaceae) Sabja

Medicine

(*i*) Extract of leaves, about 2-3 drops, are dropped into ears to treat ear-ache.

Ethnoveterinary Medicine

(*i*) Extract of leaves, about 10 ml, is poured into ears to prevent pus-formation.

150. *Ocimum tenuiflorum* L. (Lamiaceae) Tulasi

Medicine

(*i*) Leaves are powered. It is mixed in water. This paste is applied on injury caused due to axima.

(*ii*) Leaf juice (2-3 drops) is dropped into ears to cure ear-ache.

(*iii*) Extract of leaves about, two spoonful, is drunk for three days to treat stomach-ache.

Ethnoveterinary Medicine

(*i*) Leaves are crushed in whey. This paste applied on fractured organ of the pet animals.

151. *Opuntia stricta* Haw. (Cactaceae) Nivdung

Medicine

(*i*) Ripe fruits are burnt. Two or three are consumed daily for 3-4 days to cure cough.

152. *Ougeinia oojenensis* (Roxb.) Hochr. (Fabaceae) Tiwas

Miscellaneous Uses

(*i*) Wood is useful for making bullock- carts.

(*ii*) Wood is used for house construction and agricultural tools.

(*iii*) Leaves constitute good fodder for pet animals.

153. *Pandanus fascicularis* Lamk. (Pandanaceae) Kewada

Miscellaneous Uses

(*i*) Fibres are extracted from leaves and used in making of small baskets.

154. *Papaver somniferum* L. (Papaveraceae) Khas-khas

Medicine

(*i*) Seeds are crushed in water. About 2-3 drops are administered to infants in case of excessive weeping.

155. *Pedilanthus tithymaloides* (L.) Poit. (Euphorbiaceae)

Medicine

(*i*) Young leaves (3-4) are consumed with one clove to treat piles. It is followed till cure.

156. *Pergularia daemia* (Forsk.) Chiov. (Asclepiadaceae) Utran

Ethnoveterinary Medicine

(*i*) Leaves are crushed with groundnut oil. The paste is applied in case of bone facture and tied using wooden plates.

157. *Phoenix dactylifera* L. (Arecaceae) Kharik

Medicine

(*i*) Two to three seeds are boiled in milk. It is drunk for seven days as aphrodisiac.

(*ii*) Two to three dried date fruits are kept overnight in a cup of milk. The milk is drunk to treat chronic fever till cure.

158. *Physalis minima* L. (Solanaceae) Dokephodi

Ethnoveterinary Medicine

(*i*) Entire plant is crushed it is homogenized with fruit powder (3-4 fruits) of *Cucumis callosus* (Rottl.) Cogn. This paste, about 400-500 gm, is fed to cows and bullocks suffering from dysentery.

159. ***Piper betle*** **L. (Piperaceae) Nagin –pan**

Medicine

(*i*) Leaves are crushed along with 2-3 flowers of *Ricinus communis* L. It is advised twice a day for two days.

160. ***Piper nigrum*** **L. (Piperaceae) Kali-miri**

Medicine

(*i*) Seeds are crushed and powdered. It is applied on suppurating tumours or boils.

161. ***Pithecellobium dulce*** **(Roxb.) Benth. (Mimosaceae) Amali-chinch**

Medicine

(*i*) Seed paste is applied on joint to cure joint-ache for 4-5 days.

Miscellaneous Uses

(*i*) Leaves are used as fodder for goats.

(*ii*) Dried branches and wood as fuel.

162. ***Plumeria alba*** **L. (Apocynaceae) Chapha**

Medicine

(*i*) The unripe fruits are advised to smell after snake bite.

163. ***Polyanthes tuberosa*** **Linn. Amaryllidaceae Rajanigandha**

Medicine

(*i*) Stem is powdered. About spoonful of it is taken orally for seven days against general weakness.

(*ii*) Extract of leaves mixed in its stem powder is taken orally for seven days against leucorrhoea. A spoonful of it is advised per day.

164. ***Portulaca quadrifida*** **L. (Portulacaceae) Chidichi bhaji**

Medicine

(*i*) Roots are boiled in water. This decoction, 10-15 ml, is drunk to cure hepatitis till cure.

Miscellaneous Uses

(*i*) It is as a fodder for domestic animals.

165. *Prosopis cineraria* (L.) Druce (Mimosaceae) Saundad

Miscellaneous Uses

(*i*) Wood is used as fuel.

(*ii*) Wood is employed as pillars of hut.

(*iii*) Leaves and green pods are good fodder for goats.

166. *Psidium guajava* L. (Myrtaceae) Jam, Peru

Medicine

(*i*) Leaves are boiled in water. This decoction is used to gargle. This helps to reduce tooth-ache. It is followed for 2-3 days.

167. *Punica granatum* L. (Punicaceae) Dalimb

Medicine

(*i*) Young leaves are chewed to cure mouth-ulcers. It is practiced till cure.

Ethnoveterinary Medicine

(*i*) Raw fruits (4-5) are fed to pet animals for few days to deworm intestinal worms.

168. *Pupalia lappacea* (L.) Juss. (Amaranthaceae) Bend-gawat

Ethnoveterinary Medicine

(*i*) Leaves (about 200 gm) are crushed along with cumin seeds (25 gm) [*Trachyspermum ammni* (L.) Sprague]. It is fed to treat flatulence daily once till cure.

169. *Raphanus sativus* L. (Brassicaceae) Mula

Medicine

(*i*) Cooked leaves are consumed once a day to treat fever.

(*ii*) Roots are consumed to regularize digestive complaints.

170. *Rhus sinuata* Thunb. (Anacardiaceae) Amoni

Miscellaneous Uses

(*i*) Fruits are sore and edible. These are cherished by boys.

(*ii*) Leaves are fed to goats and cattle.

171. *Rhynchosia minima* DC. (Fabaceae)

Miscellaneous Uses

(*i*) Plants are useful as fodder for goats.

172. *Ricinus communis* L. (Euphorbiaceae) Erandi, Erand

Medicine

(*i*) Extract of young leaves prepared in cow-milk, about a cup, is drunk for 3-4 days to treat hepatitis.

(*ii*) Root extract is mixed in whey. A cup of it is administered per day to a patient suffering from urinary stone. It is followed for a week.

(*iii*) Extract of leaves is mixed with curd. A cup of it is taken orally at morning for few days against hepatitis.

(*iv*) Flowers are consumed with betel (*Piper betle* L.) leaf for seven days for abortion.

(*v*) Leaf lamina, collected from the leaf bent down naturally on the tree, is smoked to treat migraine.

Ethnoveterinary Medicine

(*i*) Leaves are crushed in whey. This paste is tied on fractured bone.

Miscellaneous Uses

(*i*) Dried plants are used as fuel.

173. *Rosa canina* L. (Rosaceae) Lal-Gulab

Medicine

(*i*) Flowers are crushed in water. This extract, about 5-10 ml, is drunk at morning for few days to reduce body heat.

174. *Rumex vesicarius* L. (Polygonaceae) Amberchuka

Medicine

(*i*) Leaves are cooked with leaves of fenugreek (*Trigonella foenum-graecum* L.). It is consumed to treat stomach-ache twice a day.

175. *Saccharum officinarum* L. (Poaceae) Kala-oos

Medicine

(*i*) Juice obtained from sugarcane plants (with blackish stem) is drunk, about 250 ml, for 2-3 days against acidity.

(*ii*) Juice obtained from blackish stem is drunk, about a cup daily, to regulate digestive system.

Miscellaneous Uses

(*i*) Dried leaves are used for thatching huts.

(*ii*) Dried leaves are cut into pieces and mixed in mud while making bricks.

(*iii*) Green leaves are used as fodder for domestic animals.

176. *Salvadora persica* L. (Salvadoraceae) Pilu-kathar

Medicine

(*i*) Decoction of leaves with common salt is applied on joints to reduce rheumatic pains. It is advised for few days.

(*ii*) Leaves are crushed in juice of lemon (*Citrus aurantifolia* Sw.). This paste is massaged after bursting of boils.

Ethnoveterinary Medicine

(*i*) Crushed leaves are heated. This hot paste is applied on fractured organ of pet animal.

Miscellaneous Uses

(*i*) Roots are used as tooth-brush especially by Mohamedan people. They believe that their 'Namaz (praying)' is accepted by God.

177. *Santalum album* L. (Santalaceae) Chandan

Medicine

(*i*) Leaf juice about spoonful is taken orally, twice a day for three days, to check fever.

Miscellaneous Uses

(*i*) Wood is used for making a casket for keeping especially 'kumkum' (red powder).

178. *Sapindus emarginatus* Vahl (Sapindaceae) Ritha

Medicine

(*i*) Fruit is rubbed in water. The paste is applied on toes to treat thick skin developed on toes.

Miscellaneous Uses

(*i*) Dried leaves are dipped in water for a longer period. The water is used as detergent to wash cloth.

(*ii*) Fruits are used for washing hair.

179. *Saraca asoca* (Roxb.) de wilde (Caesalpiniaceae) Ashok

Medicine

(*i*) Stem bark is powdered. About 10 gm per day is consumed with butter to cure leucorrhoea. It is advised till cure.

Miscellaneous Uses

(*i*) Wood is employed as pillars of houses and huts.

180. *Securinega leucopyrus* (Willd.) Muel.-Arg. (Euphorbiaceae) Pandhari-phuli

Miscellaneous Uses

(*i*) Leaves are fed to goats.

181. *Semecarpus anacardium L.f.* (Anacardiaceae) Biba, Bhilava

Medicine

(*i*) Extract of seed is applied on foot-cracks till cure.

(*ii*) Two to three seeds are boiled in a cup of milk. It is drunk twice a day to cure cough till cure.

(*iii*) Fruit is crushed with juice of onion (*Allium sativum* L.). It is given orally twice a day for three days against stomach-ache.

Ethnoveterinary Medicine

(*i*) One seed is crushed and homogenized with butter. It is advised twice a day to cure 'lalkhurgati' i.e. FMD (Foot and mouth disease) practiced till cure.

182. *Sesamum orientale* L. (Pedaliaceae) Til

Medicine

(*i*) Seed oil is heated and massaged to reduce body swellings and also for rheumatism. It is advised till cure.

183. *Sesbania grandiflora* (L.) Poir. (Fabaceae) Hadga

Medicine

(*i*) Extract of leaves, about 2-3 drops, are dropped into ears to prevent pus formation.

(*ii*) Extract of flowers is drunk, about a cup twice a day, to treat hepatitis. It is followed till cure.

Miscellaneous Uses

(*i*) Wood is used as a fuel.

184. *Setaria verticillata* (L.) P. Beauv. (Poaceae) Gawat

Miscellaneous Uses

(*i*) Plants are used as fodder.

(*ii*) Dried plants are used for thatching huts.

185. *Solanum incanum* L. (Solanaceae) Dorle-wange

Medicine

(*i*) Seeds are burnt; fumes are taken in mouth for 2-3 days to cure tooth-ache. This helps to deworm the tooth.

186. *Solanum melongena* L. (Solanaceae) Wange

Medicine

(*i*) Fruit is baked. It is homogenized with some asafoetida. It is applied on wounds caused due to guinea worms.

187. *Solanum virginianum* L. (Solanaceae) Bhuiringani

Medicine

(*i*) Extract of leaves is mixed with animal ghee in equal proportion. Three to four spoons taken orally twice a day for consecutive seven days to treat piles.

188. *Sonchus oleraceus* L. (Asteraceae) Mhatara

Miscellaneous Uses

(*i*) Plants are useful as fodder for cattle.

189. *Spilanthes calva* DC. (Asteraceae) Akkalkadha

Medicine

(*i*) Seeds are consumed to improve stammering till cure.

190. *Syzgygium aromaticum* Merr. and Perry (Myrtaceae) Lawang

Medicine

(*i*) Crushed flowers are placed near gums to cure tooth-ache.

191. *Syzygium cumini* (L.) Skeels (Myrtaceae) Jambul

Medicine

(*i*) Fruit is consumed for seven days against constipation.

(*ii*) Ripe fruits are consumed for few days to treat tooth-ache.

(*iii*) Leaves are dried and powdered. About spoonful of it is consumed twice a day for seven days to treat leucorrhoea.

192. *Tabernaemontana divaricata* L. (Apocynaceae) Chandani

Ethnoveterinary Medicine

(*i*) Leaves are crushed with red lead and applied on shoulder of bullocks to cure shoulder injury.

193. *Tagetes erecta* L. (Asteraceae) Zendu

Medicine

(*i*) A cup of leaf extract is taken orally at morning for 10-15 days to control diabetes.

(*ii*) Extract of leaves, about a cup, is drunk at early morning with two spoonful of animal ghee against piles. It is followed for seven days.

(*iii*) Extract of leaves, about 2-3 drops, is poured into ears to cure ear-ache till cure.

194. *Tamarindus indica* L. (Ceasalpiniaceae)Chinch

Medicine

(*i*) Seeds are powdered. Its paste prepared in water is applied on eyes to reduce redness of eyes. It is practiced till cure.

Ethnoveterinary Medicine

(*i*) Leaves are boiled with soil obtained from ant-hill. It is applied in case of bone fracture of domestic animals. It is followed for 10-15 days.

Miscellaneous Uses

(*i*) Dried branches and wood are used as fuel.

(*ii*) Timber is used in making partition wall of huts.

195. *Tectona grandis* L. (Verbanaceae) Sag

Medicine

(*i*) Seeds are rubbed in water. This paste is applied on abdominal region to prevent excessive urination.

196. *Terminalia arjuna* (Roxb.) Wight and Arn. (Combretaceae) Arjun

Medicine

(*i*) Stem bark is powdered. Its paste is prepared using coconut oil. It is applied in injuries till cure.

Miscellaneous Uses

(*i*) Wood is used for making cots.

(*ii*) Wood is employed as pillars while constructing houses.

197. *Terminalia bellirica* (Gaertn.) Roxb. (Combretaceae) Behada

Medicine

(*i*) Fruit wall is powdered. Pellets are prepared using goat-urine. One pellet is taken twice a day to treat asthma. It is followed till cure.

Miscellaneous Uses

(*i*) Leaves are used to prepare dining plates and dishes.

(*ii*) Wood is used as fuel.

(*iii*) Green leaves are fed to cattle and goats.

198. *Terminalia catappa* L. (Combretaceae) Badam, Desi-badam

Medicine

(*i*) Flowers are powdered. About spoonful of it is taken orally in honey twice a day for seven days to treat piles.

(*ii*) Leaves are heated and tied on joints to treat joint-ache.

Miscellaneous Uses

(*i*) Wood is used as fuel.

(*ii*) Wood is employed as pillar of huts.

199. *Terminalia chebula* Retz. (Combretaceae) Hirda

Medicine

(*i*) Fruits are powdered. It is boiled in water. About spoonful of it is taken orally at night to treat constipation.

(*ii*) Decoction of fruits is gargled once a day to cure mouth-ulcer.

(*iii*) Fruit are rubbed in whey. It is gargled to cure tooth-ache.

Miscellaneous Uses

(*i*) Fruit powder is mixed in coconut oil. It is applied while massaging.

200. *Tinospora cordifolia* (Wild.) Miers. (Merispermaceae) Gulwel

Medicine

(*i*) Stem pieces are boiled in water. A cup of this decoction is drunk to treat body-ache.

(*ii*) Stem is heated and crushed. The paste is prepared in coconut oil and applied on eczema.

201. *Trachyspermum ammni* (L.) Spr. (Apiaceae) Owa

Medicine

(*i*) Decoction of seeds, about spoonful per day, is taken orally to treat urinary stone till cure.

(*ii*) A spoonful of seeds are consumed against flatulence.

202. ***Trema orientalis*** **(L.) Bl. (Ulmaceae)**

Miscellaneous Uses

(*i*) Leaves are used as fodder for cattle.

203. ***Tribulus lanuginosus*** **L. (Zygophyllaceae) Gokharu**

Medicine

(*i*) Crushed roots and leaves together in are consumed against snake-bite.

204. ***Tridax procumbens*** **L. (Asteraceae) Tantani, Ekdandi**

Medicine

(*i*) Leaf paste is applied on injury to avoid septic.

205. ***Trigonella foenum-graecum*** **L. (Fabaceae) Methi**

Medicine

(*i*) Extract of leaves, about half cup is drunk twice a day for 34 days to check blood-pressure.

(*ii*) Seeds are rubbed in water. This water is used for hair washing to control dandruff and for softness of hair.

(*iii*) Entire plants are boiled in water. About a cup of this decoction is administered to treat joint-ache till cure.

206. ***Triticum aestivum*** **L. (Poaceae) Gahu**

Medicine

(*i*) Extract of entire young plants, about a cup, is consumed for seven days at early morning to treat fever and acidity.

207. ***Vernonia cinerea*** **(L.) Less. (Asteraceae) Budhhi**

Ethnoveterinary Medicine

(*i*) Young leaves are crushed. About a litre of extract is administered to bullocks and cows suffering from cough.

208. ***Vigna mungo*** **(L.) Hepper (Fabaceae) Udid**

Medicine

(*i*) Seeds are chewed. The paste is wrapped on fingers to treat whitlow.

209. *Vitis vinifera* L. (Vitaceae) Manukhe, Drakshe

Medicine

(*i*) Raisins are dipped in wine prepared by using flowers of *Madhuca longifolia* (Koen.) Macbr. Three to four raisins are consumed twice a day against asthma till cure.

(*ii*) Roots are rubbed in water. It is drunk, about half cup, as antidose for snake bite.

210. *Wattakaka volubilis* (L.*f.*) Stapf. (Asclepiadaceae) Bokadwel

Miscellaneous Uses

(*i*) Leaves and fruits are fed to goats.

211. *Withania somnifera* (L.) Dunal (Solanaceae) Ashwagandha

Medicine

(*i*) Roots are powdered. About spoonful of it taken with milk for seven days to reduce breast-ache.

Ethnoveterinary Medicine

(*i*) Roots are powdered and mixed with root extract of *Asparagus racemosus* L. About 200 gm is homogenized in one litre of water. It is administered to the buffaloes.

212. *Xanthium indicum* L. (Asteraceae) Kutri, Bindare, Kokan

Medicine

(*i*) Leaves are crushed with one bulblet of *Allium sativum* L. The extract obtained, about 1-2 drops, are poured into nostril or opposite side in case of migraine.

Miscellaneous Uses

(*i*) It is used as fuel by poor class people.

213. *Zea mays* L. (Poaceae) Maka

Medicine

(*i*) Flowers are smoked to cure asthma. This ash is dissolved in water and drunk for the same complaint.

214. *Zingiber officinale* Roxb. (Zingiberaceae) Adrak

Medicine

(*i*) Rhizome is powdered. It is mixed with common salt and asafoetida. This paste is applied on injuries twice a day to treat piles.

(*ii*) Extract obtained from rhizome is mixed with jaggery and diluted in water. Few drops of it are poured into nose to cure cold. It is advised twice a day.

215. *Zizipus glabrata* Heyne ex Roth (Rhamnaceae) Bor

Medicine

(*i*) Leaf paste is applied on the injuries. This is practiced till cure.

Miscellaneous Uses

(*i*) Stem wood is used in making agricultural equipments and handles of axe.

(*ii*) Branches being spiny and stout are used for fencing farmyards.

(*iii*) Leaves are used as fodder especially for goats.

SECTION–II
[*Shegaon, Sangrampur and Jalgaon (Jamod) Tehsils*]

1. ***Abelmoschus manihot*** **(L.) Medic. (Malvaceae) Ranbhendi**

Medicine

(*i*) Root juice, two spoonful with a cup of curd is consumed twice a day for a week to cure piles.

2. ***Abrus precatorius*** **L. (Fabaceae) Pandhari-gunj**

Medicine

(*i*) Roots are crushed in water. The extract obtained is dropped, about two-three drops, in opposite side nostril to treat migraine before sleeping at night. It is followed till cure.

3. ***Acacia leucophloea*** **(Roxb.) Willd. (Mimosaceae) Hiwwar**

Medicine

(*i*) Paste of leaves or roots is applied on injuries of after scorpion sting or due to sickle. It is also applied for sweat-blisters and prickly-heat.

4. *Acacia nilotica* (L.) Willd. ex Del. subsp. *indica* (Benth.) Brenan (Mimosaceae) Babhul, Telibabhul

Medicine

(*i*) Stem is chewed to increase strength of gums.

(*ii*) Fruit, leaves and gum in equal amount are powdered. It is homogenized with animal ghee. One spoonful of it, twice a day, is consumed to improve sperm count.

(*iii*) Stems are debarked first and then dipped in water overnight. This infusion is used to gargle in case of mouth ulcer. It is followed till cure.

(*iv*) Leaf juice, one spoonful, is mixed with two spoonful of animal ghee. This is consumed twice a day for three days to kill intestinal worms.

Miscellaneous Uses

(*i*) Wood is used for all types of farming implements and instruments.

(*ii*) Leaves and green pods are fed to goat.

5. *Acalypha malabarica* Muell.-Arg. (Euphorbiaceae)

Miscellaneous Uses

(*i*) Plants are used as fodder for cattle.

6. *Acanthospermum hispidium* DC. (Asteraceae) Gokharu

Medicine

(*i*) Seeds are mixed with the seed of *Vitex negundo* L., seed of *Mucuna pruriens* (L.) DC. Prain, roots of *Withania somnifera* (L.) Dunal and fruits of *Emblica officinalis* Gaertn. in equal amount. This powder, one spoonful with milk, is consumed. It is useful for body strengthening.

(*ii*) Seeds are crushed with the rhizomes of *Zingiber officinale* Rosc. It is then boiled in water. This decoction, one cup, is drunk before meal to cure kidney-stone. It is followed till cure.

7. *Achyranthes aspera* L. (Amranthaceae) Apamarg, Aghada, Khatara

Medicine

(*i*) The whole plant is burnt and ash is dissolved in water. It is boiled and salt is obtained. It is then applied on injuries of anus caused due to piles and fissures etc. till cure.

(*ii*) Entire plants are dried in shadow. It is powdered. One spoonful powder with honey is consumed at morning for seven days to cure asthma.

Ethnoveterinary Medicine

(*i*) Roots are crushed with oil cake to improve milk quality and quantity of cows and she-buffaloes.

8. *Aegle marmelos* (L.) Corr. (Rutaceae) Bel

Medicine

(*i*) Fruit is homogenized with honey. One spoonful is consumed for colic. It is followed till cure.

Miscellaneous Uses

(*i*) Stem is used to prepare a 'Chaurang', a four-legged wooden artifact used on some auspicious or religious ceremony.

9. *Aerva lanata* (L.) Juss. ex Schult. (Amaranthaceae)

Miscellaneous Uses

(*i*) Plants are used as fodder for cattle.

10. *Ageratum conyzoides* L. (Asteraceae)

Miscellaneous Uses

(*i*) It is used as fodder for cattle.

11. *Allium cepa* L. (Liliaceae) Kanda

Medicine

(*i*) Bulbs are halved and heated. Sweat-blisters are wormed with the hot bulb to cure it.

(*ii*) Inflorescence axis is crushed. Few drops (two-three) are poured into ears to treat ear-ache and pus formation.

12. *Allium sativum* L. (Liliaceae) Lasoon

Medicine

(*i*) Few bulblets are crushed with roasted gram (*Cicer arientinum* L.) at night daily for clear voice.

(*ii*) Crushed bulblets are hold in teeth to kill germs in mouth.

(*iii*) Bulblets are crushed and mixed with camphor (*Ocimum kilimandscharium* Guercke.) and powder of leaves of *Nicotiana tabacum* L. in equal amount. This paste is applied on forehead to cure migraine.

(*iv*) Bulb juice is applied on uvula to cure whooping cough. Twice or thrice a day.

13. *Aloe vera* L. (Liliaceae) Korphad

Medicine

(*i*) Latex obtained from leaves is warmed and then dropped into ears to prevent pus formation during ear-ache.

14. *Alternanthera tenella* Colla (Amaranthaceae)

Miscellaneous Uses

(*i*) The plants are used as fodder for cattle and goat.

15. *Alysicarpus tetragonolobus* Edgew. (Fabaceae) Gavat

Miscellaneous Uses

(*i*) It is used as fodder for cattle.

16. *Amaranthus polygonoides* L. (Amranthaceae) Tandulja

Miscellaneous Uses

(*i*) Plants are used as fodder for cattle.

(*ii*) Leaves are used as vegetable.

17. *Amaranthus viridis* L. (Amaranthaceae) Matla

Miscellaneous Uses

(*i*) Plants are used as fodder for cattle.

(*ii*) Leaves are used as vegetable.

18. *Ampelocissus latifolia* (Roxb.) Planch. (Vitaceae) Ran-draksha

Medicine

(*i*) Roots are dried in shadow and finely powdered. Half tablespoon powder is mixed in curd. This is consumed for three days at the time of sunrise to stop blood dysentery.

19. *Annona reticulata* L. (Annonaceae) Ramphal

Medicine

(*i*) Pulp of fruit is mixed with the decoction obtained from the leaves of *Tinospora cordifolia* (Willd.) Miers., stem paste of *Glycyrrhiza glabra* L., seeds of a *Elettaria cardamomum* L. and candy in equal amount. It is consumed twice a day for seven days to avoid excess body heat. Oily, sweet and sour diet is to be avoided.

20. *Annona squamosa* L. (Annonaceae) Sitaphal

Medicine

(*i*) Seed powder is mixed with jaggery (*Saccharum officinarum* L.). A spoonful of it is consumed to kill intestinal worms. It is consumed for seven days before meal.

21. *Argemone mexicana* L. (Papaveraceae) Kate-bilayat, Bilayat

Medicine

(*i*) Latex of the plant is rubbed in case of scabies, eczema and cracks on sole of foot.

(*ii*) Latex is applied on sweet-lump (Gham-bend) to cure it. It is advised till cure.

(*iii*) Oil is extracted from seed. It is applied on any injury to cure it.

(*iv*) Latex is also applied on common injuries.

22. *Aristolochia bracteolata* Lamk. (Aristolochiaceae) Gindiyan

Medicine

(*i*) Two-three drops of leaf juice are dropped into ears to treat ear-ache.

(*ii*) Leaf juice is mixed with lump-sugar. A spoonful is consumed at morning for twenty-one days to increase body strength.

23. *Azadirachta indica* A. Juss. (Meliaceae) Kadu-limb

Medicine

(*i*) Leaf juice is applied on eyes for inflammation.

(*ii*) Flowers especially in the month of 'Chaitra' are rubbed on body to repel snakes and scorpion.

24. *Balanites aegyptiaca* (L.) Del. (Balanitaceae) Hingne

Medicine

(*i*) Leaves are chewed and then the saliva obtained is applied on eyes for eye injury.

(*ii*) Leaves are rubbed in animal ghee to prepare paste. This paste is applied on foot hill-ache till cure.

25. *Barleria prionitis* L. (Acanthaceae) Katya-krotya, Kate-karwanti

Medicine

(*i*) Leaves are chewed and then gargled with whey to cure mouth-ulcer.

26. *Bauhinia racemosa* Lamk. (Caesalpiniaceae) Apta

Miscellaneous Uses

(*i*) Leaves and green pods used as fodder for cattle.

27. *Biophytum sensitivum* (L.) DC. (Oxalidaceae) Lajalu

Medicine

(*i*) Roots are tied around waist to avoid different diseases. This is followed during especially during the period of 'Uttara Nakshatra'.

28. *Blepharis repens* (Vahl) Roth (Acanthaceae) Hadsan

Medicine

(*i*) Leaves are consumed alongwith 'Chapati' of *Triticum aestivum* L. to cure the bone fracture. It is consumed for about a week.

29. *Boerhavia diffusa* L. (Nyctaginaceae) Punarwel, Punar-nava, Sathi, Ghotuli

Medicine

(*i*) Plant juice with sugar is consumed to maintain youthness.

(*ii*) Entire plants are dried and powdered. It is mixed in cow's ghee. It is stored for one day in red bottle and placed in sunlight. It is then mixed in cow's milk. This is applied on swollen body parts.

Miscellaneous Uses

(*i*) Plant is used as fodder for cattle.

30. *Boerhavia erecta* L. (Nyctaginaceae) Punarun

Miscellaneous Uses

(*i*) It is fed to goats.

31. *Boerhavia repens* L. (Nyctaginaceae)

Miscellaneous Uses

(*i*) Leaves are used as fodder for cattle especially for goats and sheeps.

32. *Brassica juncea* (L.) Czern. and Coss. (Brassicaceae) Mohri

Medicine

(*i*) Seed oil is mixed in the paste of *Cuscuta chinensis* Lamk. This paste is applied on leucoderma.

33. *Butea monosperma* (Lamk.) Taub. (Fabaceae) Palas

Medicine

(*i*) Flowers are soaked in water contained in earthern pot. The flowers are then crushed and boiled in the same water. This decoction, one cup a day for a fortnight, is drunk to cure various urinary complaints

(*ii*) Seeds are rubbed in oil. This paste is applied on scabies and eczema.

Miscellaneous Uses

(*i*) Leaves are used for thatching roof.

(*ii*) They are also employed for dining plates.

34. *Caesalpinia bonduc* (L.) Roxb. (Caesalpiniaceae) Sagargoti

Medicine

(*i*) Decoction of leaves or its extract is drunk to treat rheumatism. A cup of it is administered for seven to eight days.

(*ii*) Leaves are dried and finely powdered. It is consumed with one spoonful of water at morning before meal to cure piles.

(*iii*) Leaves are dried in moon-light and powdered. One spoonful of it is mixed in a cup of whey. It is consumed for five days. This is administered to treat excessive bleeding during menstruation.

(*iv*) Fruits are boiled in water. Decoction is drunk for three days to cure cough in case of children.

(*v*) Kernels are crushed and mixed with jaggery (*Saccharum officinarum* L.) and pellets are prepared. Two pellets a day are advised to treat colic.

35. *Caesalpinia decapetala* (Roth) Alst. (Caesalpiniaceae) Chillari

Miscellaneous Uses

(*i*) They are planted as live-edge around cultivated fields and farmyards.

36. *Cajanus cajan* (L.) Millsp. (Fabaceae) Tur

Miscellaneous Uses

(*i*) Stems are employed for partition wall of huts.

(*ii*) They are also used to prepare baskets.

37. *Calotropis gigantea* (L.) R. Br. (Asclepiadaceae) Pandhari-rui

Medicine

(*i*) Flower buds of this plant along with leaves of *Justicia adhatoda* L., seeds of *Piper nigrum* L., rhizome of *Zingiber officinale* Rosc. are crushed finely. Pellets are then prepared from it. One pellet per day is consumed for seven days after sunset by the patient suffering from knee-ache.

(*ii*) Flower juice (50 ml), latex (20 ml) of this plant is mixed with cow's ghee (100 ml). It is boiled slightly till the latex and juice coagulate. Ten drops of this are mixed in one cup of cow's milk. It is drunk twice a day for a week to improve sperm production.

(*iii*) Flowers are dried in shadow and consumed along with cow's ghee daily for a fortnight to cure impotency.

(*iv*) The dried leaves are burnt with oil. The ash is collected on copper pot. This ash is then used to cure various skin diseases.

(*v*) The latex is spread on candy and consumed during winter season to cure the patient suffering from mental derangement.

(*vi*) Dried flowers and seeds of *Piper nigrum* L. are powdered and mixed in jaggery (*Saccharum officinarum* L.). Pellets of one gram are prepared. Two pellets are consumed twice a day for twenty-four days to cure asthma.

Ethnoveterinary Medicine

(*i*) Latex is applied on wounds to kill worms.

38. *Calotropis procera* (Ait.) R. Br. (Asclepiadaceae) Rui

Medicine

(*i*) Latex is applied on scorpion string.

(*ii*) Roots are levigated or boiled in water. This water is added in bath water to treat bloating.

(*iii*) Leaves are warmed with oil. It is quickly wrapped around knees for treating arthritis. This is followed till cure.

Ethnoveterinary Medicine

(*i*) Leaves are fed to animals suffering from colic.

(*ii*) Leaves are also fed to animals and then oil is administered orally as a remedy against flatulence.

(*iii*) Latex is mixed with hen-egg and then applied on leg to treat twisting.

39. *Capparis grandis* L. (Capparidaceae) Arjun-bor

Miscellaneous Uses

(*i*) Leaves are used as fodder for cattle and goats.

40. *Capparis sepiaria* L. (Capparidaceae) Dakin-bor

Miscellaneous Uses

(*i*) Leaves are used as fodder for goats.

41. *Capparis zeylanica* L. (Capparidaceae) Waghuti, Waghati

Medicine

(*i*) Leaves of this plant and of *Justicia adhatoda* L. are boiled. They are tied on tumor till cure.

42. *Carica papaya* L. (Caricaceae) Papai

Medicine

(*i*) Two seeds are powdered. It is mixed in spoonful water. This infusion is administered to the kids to kill intestinal worms. Depending on the age of kids the number of seeds is increased.

43. *Carissa congesta* Wight (Apocynaceae) Karwand, Dongrachi-kali-maina

Medicine

(*i*) Fruits are consumed along with common salt once a day to kill intestinal worms.

Miscellaneous Uses

(*i*) Ripe fruits are cherished by all.

44. *Cassia auriculata* L. (Caesalpiniaceae) Awali

Medicine

(*i*) Seeds of this plant and coupted grains of *Cicer arientinum* L. About one full fist is consumed to cure knee-ache, till cure.

(*ii*) Leaves and common salt are tightly tied on twisted body part till cure.

(*iii*) Leaf juice is applied on abdomen to lower body heat and to treat sunstroke.

45. *Cassia fistula* L. (Caesalpiniaceae) Bhawa

Medicine

(*i*) Fruit pulp mixed with honey about 10 gm is administered to check dysentery.

46. *Cassia obtusifolia* L. (Caesalpiniaceae) Tarota

Medicine

(*i*) Leaves are chewed and saliva is gargled at morning to cure ulcerated mouth.

47. *Cassine alberns* (Retz.) Kosterm. (Celastraceae) Bhutakhet

Medicine

(*i*) Dried roots are powdered and boiled in cow's urine. It is applied on head at night to reduce fever.

(*ii*) Leaf juice is applied on injury and also its paste is used for piles.

48. *Cayratia trifolia* L. (Vitaceae) Tivnya

Miscellaneous Uses

(*i*) Plant body is used as fodder for goats.

49. *Celosia argentea* L. (Amaranthaceae) Shahamendha, Kurudu, Kombda

Medicine

(*i*) Roots are chewed to increase strength of gums.

(*ii*) Roots are boiled in water. They are extracted. Two spoonful of extract, twice a day, for three to four days is advised to patients suffering from jaundice.

50. *Centella asiatica* (L.) Urb. (Apiaceae) Bramhi, Bophali

Medicine

(*i*) Dried fine powder of whole plants is mixed in cow's ghee. It is boiled slightly and then filtered. It is stored in a red bottle for one day in sunlight. Ten drops of it in one cup of cow's fresh milk is given twice a day for a month to avoid mental regardness and in case of in articulation. It is also used as a brain tonic.

(*ii*) Leaf juice is mixed in seed powder of *Cuminum cyminum* L. This paste is applied on stomach to remove body heat.

51. *Cestrum nocturnum* L. (Solanaceae) Rat-Rani

Medicine

(*i*) Flowers are homogenized with honey and a piece of stem of *Glycyrrhiza glabra* L. A spoonful of it is consumed at morning before meal to cure colic.

52. *Chlorophytum borivilianum* Santapau and Fernand. (Liliaceae) Safed Musali

Medicine

(*i*) A spoonful powder of dried root tubers in a cup of milk is consumed as tonic. This is followed twice a day for a month during winter season.

53. *Cicer arietinum* L. (Fabaceae) Harbhara

Medicine

(*i*) Seeds of this plant, jaggery (*Saccharum officinarum* L.) and roots of *Nardostachys jatamansi* DC. are mixed thoroughly. This is applied on forehead to treat migraine till cure.

54. *Cissampelos pareira* L. (Menispermaceae) Pandhra-vasu

Medicine

(*i*) Roots are cooked with rice. Cow's milk is added in it. It is consumed thrice a day to cure jaundice till cure.

(*ii*) Twenty one pieces of roots are tied in black thread. These are tied on heap at new moon day to cure jaundice.

Miscellaneous Uses

(*i*) Plants are used as fodder for cattle and especially for goats.

55. *Cissus quadrangula* L. (Vitaceae) Kandyawel

Medicine

(*i*) Stems are crushed and paste is prepared. This paste is applied on dislocation of joint of fingers.

56. *Citrullus colocynthis* (L.) Schrad. (Cucurbitaceae) Kadu-indrawan

Medicine

(*i*) Stems are dried and powdered. It is homogenized with honey and pellets are prepared. Two pellets are advised daily for three days to cure ulcer.

57. *Citrus aurantifolia* (Christm.) Sw. (Rutaceae) Limbu

Medicine

(*i*) Fruit juice is added in ten gm seed powder of *Piper nigrum* L., ten gm of ammonium chloride and ten gm of sulphur. It is homogenized. Then leaf juice of *Azadirachta indica* A. Juss. is also added to prepare paste. This paste is applied to cure scabies.

(*ii*) Fruit is sliced. On each slice is added some turmeric powder (*Curcuma longa* L.) jaggery (*Saccharum officinarum* L.) and vermilion. Then it is cooked and applied on piles.

(*iii*) A quarter of fruit, some common salt, powder of rhizome of *Zingiber officinale* Rosc., powder of seeds of *Piper nigrum* L. and some sugar kept together in air tight container for twelve hours. This mixture is homogenized and consumed before meal at morning to cure acidity. This is advised for three days.

58. *Citrus sinensis* L. (Osbeck) (Rutaceae) Mosambi

Ethnoveterinary Medicine

(*i*) Juice of fruit bark is applied on hand and then the prolaps of animal is pushed inside.

59. *Cleome gynandra* L. (Capparidaceae) Pandhari-Tiwan

Medicine

(*i*) Leaves are rubbed on forehead to cure migraine.

(*ii*) Few drops of leaf juice are dropped into nose for head ache.

60. *Cleome viscosa* L. (Capparidaceae) Tilwan

Medicine

(*i*) Leaf juice is dropped into nose for three days before sleeping at night to cure influenza.

61. *Clerodendrum multiflorum* (Burm. *f.*) O. Ktze. (Verbenaceae) Arni

Miscellaneous Uses

(*i*) Leaves are used as fodder for goats.

62. *Cocculus hirsutus* (L.) Diels. (Menispermaceae) Wasan

Medicine

(*i*) Leaf paste is applied on injury.

(*ii*) Leaves are rubbed in water. One spoonful of it is given twice a day to cure fever till cure.

(*iii*) Leaf paste is applied on any injury to stop blood.

Miscellaneous Uses

(*i*) Leaves are used as fodder for goats.

(*ii*) The stems are used tie bundles of fodder.

63. *Cocos nucifera* L. (Arecaceae) Naral

Medicine

(*i*) Coconut shell is first burnt. The ash is dissolved in water. It is boiled and salt is obtained. It is then applied on the injuries of anus caused due to piles and fissures.

(*ii*) Coconut shell is burnt and carbon is obtained on a brass container. This carbon paste is applied on the injury to control scabies and eczema.

64. *Combretum albidum* G. Don (Combretaceae) Dev-pala

Miscellaneous Uses

(*i*) Stems are used for partition wall of huts.

(*ii*) Leaves are used as fodder for goats.

65. *Commelina benghalensis* L. (Commelinaceae) Kena

Miscellaneous Uses

(*i*) Plant is used as fodder for cattle.

66. *Commelina forskalaei* Vahl (Commelinaceae) Kenpat

Miscellaneous Uses

(*i*) Leaves are used as fodder for cattle.

67. *Convolvulus arvensis* L. (Convolvulaceae) Chandawel

Miscellaneous Uses

(*i*) Leaves are used as fodder for cattle.

68. *Corchorus depressus* L. (Tiliaceae) Harankhuri

Medicine

(*i*) Leaves are rubbed in curd and consumed for seven days before meal at morning to cure sunstroke and to check bleeding during urination.

69. *Corchorus olitorius* L. (Tiliaceae) Chuch

Miscellaneous Uses

(*i*) Leaves are used as vegetable.

70. *Corchorus trilocularis* L. (Tiliaceae) Haran-khuri

Miscellaneous Uses

(*i*) Plants are used as fodder for cattle.

71. *Cordia gharaf* (Forsk.) Eherenb. and Asch. (Ehretiaceae) Gondhan:

Ethnoveterinary Medicine

(*i*) Bark is tightly packed on the fractured part of pet animals.

Miscellaneous Uses

(*i*) Stem is used to prepare rolling pin.

(*ii*) Wood is used for preparing tambourine.

(*iii*) Leaves are used as fodder for cows, bullocks and buffaloes.

(*iv*) Ripe fruit are consumed by local people.

72. *Cryptolepis buchanani Roem* and Schult. (Periplocaceae) Mendhole

Miscellaneous Uses

(*i*) Leaves and green pods are used as fodder for goats.

73. *Cryptostegia grandiflora* R. Br. (Periploaceae)

Miscellaneous Uses

(*i*) Leaves are used as fodder especially for goats.

74. *Cucurbita maxima* Duch. ex Lam. (Cucurbitaceae) Devdangar

Medicine

(*i*) Leaves are boiled in water. It is used to bath for better health during pre-and post-delivery period.

75. *Curcuma amada* Roxb. (Zingiberaceae) Ambe-halad

Medicine

(*i*) Decoction of rhizome is drunk at morning and evening. A cup of it is advised for the patient suffering from kidney-stone till cure.

76. *Curcuma longa* L. (Zingiberaceae) Halad

Medicine

(*i*) Rhizome of this plant, seeds of *Piper nigrum* L. and rhizome of *Zingiber officinale* Rosc. are crushed finely. It is then boiled in water. This decoction about two spoonful daily is advised to increase memory and brilliancy.

77. *Cuscuta chinensis* Lamk. (Cuscutaceae) Amarwel

Medicine

(*i*) Decoction of stem, coconut oil (*Cocos nucifera* L.) and fine powder of leaves of *Lawsonia inermis* L. are mixed. This paste is applied on bald for hair growth to control hair fall and also as hair-dye.

78. *Cynadon dactylon* Pers. (Poaceae) Durwa, Haral

Medicine

(*i*) Seeds powder is homogenized with honey or mixed with candy. One spoonful is consumed daily at night to maintain youthness.

(*ii*) Seeds are soaked in cow's urine for seven days. This process is known as 'Bhavna dene'. These are dried in shadow for seven days and finely powdered with asafoetida and 'black salt'. This mixture, a spoonful in one glass of lukewarm water at night, is drunk to clear urinary problems, kidney stone and stomach problem. It is also added in milk for bile at night and before dinner for uneasiness.

(*iii*) Decoction of leaves is consumed for twenty one days at night to cure kidney-stone.

(*iv*) Roots are mixed with rhizome of *Zingiber officinale* Rosc. in 'Hemant' season(1:1 ratio by weight). It is powdered. A spoonful of it is consumed along with milk. It is used to increase strength.

Miscellaneous Use

(*i*) It is used as fodder for cattle.

79. *Cyperus rotundus* L. (Cyperaceae) Nagarmotha, Lavhali, Lau

Medicine

(*i*) Fine powder of dried root tubers is homogenized with honey. One spoonful of it is consumed per day to treat cough and cold till cure.

(*ii*) Root tubers are rubbed in water which is used as deodorant at bath.

(*iii*) Leaves are chewed against bite of 'Dhaman', an aquatic snake.

80. *Datura innoxia* Mill. (Solanaceae) Dhotra

Medicine

(*i*) Leaf juice is applied on cheek to treat mumps till cure.

(*ii*) Seed are soaked in cow's urine for twenty-four hours. These are dried in shadow. It is packed in cloth and bitten with the help of a stick to remove the seed coat of seeds. These are rubbed in cow's urine to prepare paste. This paste is applied on body part to render it anesthetic.

(*iii*) Seeds are mixed with cardamom seeds (*Elettaria cardamomum* L.) and are powdered. It is mixed in animal ghee. It is applied to cure piles.

(*iv*) Five leaves of it and leaves of *Calotropis procera* (Ait.) R.Br., and castor oil (*Ricinus communis* L.) is applied on them. These are wrapped on knees to check knee-ache.

81. *Desmodium gangeticum* (L.) DC. (Fabaceae) Chiktya

Miscellaneous Uses

(*i*) Leaves are used as fodder for cattle.

82. *Dichrostachys cinerea* (L.) Wight and Arn. (Mimosaceae) Yelatar

Medicine

(*i*) Roots are powdered. It is applied on gums to reduce toothache.

83. *Digera muricata* (L.) Mart. (Amaranthaceae) Bokadbhaji

Miscellaneous Uses

(*i*) Plants are used as fodder for cattle.

(*ii*) Leaves are used as vegetable.

84. *Diplocyclos palmatus* (L.) Jeffery (Cucurbitaceae) Shivlingi, Kali-apmari

Miscellaneous Uses

(*i*) Seeds are kept near habitations to keep snake away.

85. *Dolichandrone falcata* (Wall. ex DC.) Seem. (Bignoniaceae) Medshingi

Medicine

(*i*) Leaves are boiled in water. This hot water is used to warm body to treat rheumatism and arthritis. This water is also added in bath water to treat these diseases.

(*ii*) Leaf juice, about one cup, is drunk for seven days at morning. The patient is advised to stay out of habitations. This process is useful for patient suffering from leprosy.

86. *Echinops echinatus* Roxb. (Asteraceae) Utkttyar

Medicine

(*i*) Roots are consumed with betel leaves (*Piper betle* L.) for a fortnight to cure impotency.

87. *Eclipta prostrata* (L.) L. (Asteraceae) Maka

Ethnoveterinary Medicine

(*i*) Roots are fed to pet animals to cure injuries.

88. *Ehretia aspera* Roxb. (Ehretiaceae) Lokhandi

Ethnoveterinary Medicine

(*i*) Leaf juice is applied on injury to deworm it.

89. *Elephantopus scaber* L. (Asteraceae) Gavat

Miscellaneous Uses

(*i*) It is used as fodder for cattle.

90. *Elettaria cardamomum* L. (Zingiberaceae) Veldoda

Medicine

(*i*) Fruit powder with milk or milk cream and powder of ginger (*Zingiber officinale* Rosc.) is used for bile till cure.

91. *Emblica officinalis* Gaertn. (Euphorbiaceae) Amla, Awla

Medicine

(*i*) Dried leaves are roasted and then powdered. Paste is prepared using coconut oil (*Cocos nucifera* L.). This paste is used to cure injuries, burns and blisters.

(*ii*) Dried fruit pieces known as 'Awalakanthi' are mixed with leaves of *Ficus amplissima* J.E.Sm., seeds of *Cuminum cyminum* L. and candy. These are powdered. A spoonful powder is consumed for twenty-one days twice to cure tuberculosis and diarrhoea.

(*iii*) Leaves are chewed and gargled for mouth.

92. *Erythrina suberosa* Roxb. (Fabaceae) Bhangora

Medicine

(*i*) Entire plant is boiled with coconut oil (*Cocos nucifera* L.). It is filtered. The oil is used as hair tonic to control hair fall and to avoid dandruff.

93. *Eucalyptus camaldulensis* Dehnh. (Myrtaceae) Nilgiri

Medicine

(*i*) Flower extract is applied in teeth-cavity for curing tooth-ache.

(*ii*) Leaf juice is used as tooth-paste to cure tooth-ache. Fumes obtained from the leaves are also taken in mouth to care tooth-ache.

94. *Euphorbia indica.* Lamk. (Euphorbiaceae) Dudhi, Kawi, Jangli-chaha

Medicine

(*i*) Leaves are washed and then crushed in water. A spoonful is drunk to cure diarrhoea and dysentery.

(*ii*) Roots are boiled in water. This decoction is drunk for twenty-one days at morning. One cup is advised to cure mouth ulcer and uncerated mouth. Tea is forbidden during this treatment.

Miscellaneous Uses

(*i*) Plants are used as fodder for cattle.

95. *Euphorbia neriifolia* L. (Euphorbiaceae) Sabar-kand

Medicine

(*i*) Spines of the stem are removed. Paste of bajara flour (*Pennisetum americanum* L.) is applied on it along with some of asafoetida. It is warmed. It is extracted to obtain juice. A spoonful this juice thrice a day is advised for children to cure cough till cure.

96. *Euphorbia prunifolia* Jacq. (Euphorbiaceae) Dudhi

Miscellaneous Uses

(*i*) Leaves are used as fodder for cattle.

97. *Euphorbia tirucalli* L. (Euphorbiaceae) Kshar

Medicine

(*i*) Stem powder is mixed with jaggery (*Saccharum officinarum* L.). Pellets of equal size are prepared. Seven pellets are consumed for a week before meal to cure jaundice. Common salt and oil is prohibited during the treatment.

(*ii*) Stem pieces are consumed along with betel leaves (*Piper betle* L.) to cure piles.

(*iii*) Latex is applied on joint to cure arthritis.

(*iv*) Pieces of stem are burnt along with bull's horn. The smoke is passed onto injuries caused due to piles. This process is followed till cure.

Miscellaneous Uses

(*i*) Stem powder is spread in farm before sugarcane planting to avoid pathogenic diseases.

98. *Ficus benghalensis* L. (Moraceae) Wad

Medicine

(*i*) A newly formed dried root, about 50 gm, is rubbed in cow's milk. This is drunk in three doses to treat jaundice.

(*ii*) Newly born aerial roots are boiled in water. This decoction is applied on head as hair tonic.

(*iii*) Dried fruit powder, two spoonful of honey, one spoonful of cow's ghee and three spoonful of candy are homogenized. This mixture is administered for a patient suffering from impotency. This process is advised twice a day for twenty-one days.

(*iv*) Leaves are rubbed in oil. It is filtered and applied on the swollen joint to reduce pains till cure.

(*v*) Newly developed twenty-one roots are tied by black thread. It is tied on new moon on ankle, knees or waists etc. to reduce ache.

Ethnoveterinary Medicine

(*i*) Leaves are levigated in water. This paste is given to pet animals suffering from swollen leg.

99. *Ficus racemosa* L. (Moraceae) Umber

Medicine

(*i*) Unripe receptacles are dried, powdered and mixed in milk. One or two spoonful of it is consumed per day for a fortnight to cure impotency.

(*ii*) Leaf juice is also used as hair tonic.

(*iii*) Latex is applied on tumor.

(*iv*) Latex is mixed cow's milk and drunk it before sunrise for three days to cure jaundice.

(*v*) Latex is extracted from stem before sunrise and applied on twisted part of the body till cure.

100. *Ficus religiosa* L. (Moraceae) Pimpal

Medicine

(*i*) One spoonful of root extract is drunk before to increase red blood cells. It is advised for two to three weeks.

101. *Ficus virens* Dryand. (Moraceae) Pat-pimpali

Medicine

(*i*) Stem powder is mixed with equal quantity of ginger (*Zingiber officinale* Rosc.), nutmeg (*Myristica fragrans* Houtt.) and homogenized with honey. About five gm is consumed for good physical stamina.

102. *Flacourtia indica* (Burm. *f.*) Merr. (Flacourtiaceae) Kat-bor

Miscellaneous Uses

(*i*) Leaves are used as fodder for goats and sheeps.

103. *Gloriosa superba* L. (Liliaceae) Kal-lawi

Medicine

(*i*) A spoonful of latex is mixed with rice to treat dysentery till cure.

104. *Glycyrrhiza glabra* L. (Fabaceae) Bhunj, Jeshtha-madh

Medicine

(*i*) Powder of roots is homogenized with honey and applied in mouth to cure mouth ulcer.

(*ii*) This powder is homogenized with honey. A spoonful of it is drunk. Salt water or whey is drunk to cause vomiting. This is practiced to reduce liver problem.

105. *Gmelina arborea* Roxb. (Verbenaceae) Shivan

Miscellaneous Uses

(*i*) Leaves are used as fodder for goats.

(*ii*) Wood is used for house construction.

106. *Gomphrena globosa* L. (Amaranthaceae) Supari

Medicine

(*i*) Fruit is levitated in water and paste is prepared. This paste is applied on forehead to cure head-ache.

107. *Goniogyna hirta* (Willd.) Ali (Fabaceae)

Miscellaneous Uses

(*i*) Plants are used as fodder for cattle.

108. *Gossypium hirsutum* L. (Malvaceae) Kapus, Kapas

Medicine

(*i*) Roots are crushed in water. The extract is applied in case of scorpion sting.

Miscellaneous Uses

(*i*) Stems are employed in making of grain containers.

109. *Grewia flavescens* A. Juss. (Tiliaceae) Chopdi

Miscellaneous Uses

(*i*) Plants are used as fodder for cattle.

110. *Hardwickia binata* Roxb. (Caesalpiniaceae) Anjan

Miscellaneous Uses

(*i*) Leaves are used as fodder especially for milching, cows and buffaloes.

(*ii*) Timber is used for hut construction, agricultural implements and tools.

111. *Helicteres isora* L. (Sterculiaceae) Mural-sheng

Medicine

(*i*) Fruits are rubbed in water. This paste, about one spoonful, a day is administered to children for stomach-ache and dysentery for two-three days.

112. *Hemidesmus indicus* (L.) R. Br. (Periplocaceae) Dudhali

Miscellaneous Uses

(*i*) Leaves are used as fodder for goats.

113. *Hibiscus rosa-sinensis* L. (Malvaceae) Jaswandi, Ghanti

Medicine

(*i*) Flowers are directly consumed by the women for better body fitness.

(*ii*) Flower is dipped in coconut oil (*Cocos nucifera* L.) overnight. The oil is then used as hair tonic.

(*iii*) Leaf juice, a spoonful is taken, twice daily to cure dysentery till cure.

(*iv*) Leaf juice, about a spoonful, is given to children to cure colic till cure.

Miscellaneous Uses

(*i*) Dried stick is used for portion of huts.

114. *Holarrhena pubescens* (Buch.-Ham.) ex G. Don (Apocynaceae) Indrajav

Medicine

(*i*) Seeds are burnt and the ash is mixed with jaggery (*Saccharum officinarum* L.). Nine pellets of equal size are prepared. Three pellets for a day for killing intestinal worms are advised.

(*ii*) Seed powder, one and half spoonful, four times a day for twenty-four days is administered with water to treat acidity.

115. *Holoptelea integrifolia* (Roxb.) Planch. (Ulmaceae) Papda

Miscellaneous Uses

(*i*) Seeds are consumed by local people.

116. *Hordeum vulgare* L. (Poaceae) Jawas

Medicine

(*i*) Leaf paste is applied on fingertip near nail to cure nail bed abcess.

(*ii*) Seeds are mixed with the bulbs of *Allium sativum* L. turmeric (*Curcuma longa* L.) powder and coconut oil (*Cocos nucifera* L.). This mixture is warmed and paste is prepared. It is applied on finger tips to cure nailbed abcess.

Miscellaneous Uses

(*i*) Plants are used as fodder.

117. ***Ipomoea carnea*** **Jacq. subsp.** ***fistulosa*** **(Mart. ex Choisy) Austin (Convolvulaceae) Nagin, Besharam**

Medicine

(*i*) Animal ghee is applied on the leaves and warmed it. Then these are wrapped around injury to remove pus from it.

Miscellaneous Uses

(*i*) Stems are used to prepare mats used for bullock-carts or partition wall of huts.

118. ***Ipomoea hederifolia*** **L. (Convolvulaceae)**

Miscellaneous Uses

(*i*) Plants are used as fodder for cattle and especially for goats.

119. ***Ipomoea obscura*** **(L.) Ker.-Gawl. (Convolvulaceae) Pungalicha-pala**

Miscellaneous Uses

(*i*) Plants are used as fodder for cattle and particularly for goats.

120. ***Ipomoea quamoclit*** **L. (Convolvulaceae) Ganesh-wel**

Medicine

(*i*) Smoke of dried leaves is passed on injury of piles daily once till cure.

121. ***Justicia adhathoda*** **L. (Acanthaceae) Adulsa**

Medicine

(*i*) Decoction of leaves is homogenized with equal quantity of honey. One cup of it is consumed twice a day to cure cough, cold and fever.

(*ii*) Leaves are boiled in water. The vapours are passed on infected body part. The decoction is also rubbed on the infecting part to cure eczema.

(*iii*) Leaf juice or leaf powder is homogenized with jaggery(*Saccharum officinarum* L.). Pellets are prepared and consumed at any fixed time of a day. Two pellets for a week to cure whooping cough are advised.

(*iv*) Decoction of leaves is taken orally for twenty one days. One cup for twice a day is advised to cure asthma.

122. *Kalanchoe pinnata* (Lamk.) Pers. (Crassulaceae) Panfuti

Medicine

(*i*) Leaves are crushed and juice is obtained. Half cup of it is consumed only once to regulate menstrual cycle.

123. *Kedrostis rostrata* (Rottl.) Cogn. (Cucurbitaceae) Mirchi-kand

Medicine

(*i*) One piece of tuber is wrapped by the leaves of betel (*Piper betle*) L. It is advised to consume by the patient suffering from snake-bite. Only one dose is advised.

Ethnoveterinary Medicine

(*i*) Paste of root is applied for the pet animals suffering from liver and in case of unwillingness to eat fodder.

124. *Lagenaria siceraria* (Molina) Standl. (Cucurbitaceae) Bhopla

Medicine

(*i*) Fruits are consumed as vegetable for daily to cure sorosis till cure.

125. *Lantana camara* L. (Verbenaceae) Katmani

Medicine

(*i*) Leaves are rubbed in equal amount of jaggery (*Saccharum officinarum* L.). Seven pellets are prepared. These are consumed for seven days at morning to cure piles.

(*ii*) Leaf juice, a spoonful, is mixed in a cup of cow's milk and drunk at morning before meal to cure jaundice till cure.

Miscellaneous Uses

(*i*) Plants as used for making partition walls.

126. *Lawsonia inermis* L. (Lythraceae) Mehandi

Medicine

(*i*) Leaves are dried and powdered. Paste obtained is applied on hill crack till cure.

127. *Leucas cephalotes* (Koenig ex Roth) Spreng. (Lamiaceae) Gawat

Miscellaneous Uses

(*i*) Plants are used as fodder for cattle.

128. *Limonia acidissima* L. (Rutaceae) Kauth

Medicine

(*i*) Leaves are chewed to cure mouth ulcer.

(*ii*) Leaf juice, about a cup, is drunk at morning before food to cure asthma. It is advised for fortnight.

(*iii*) Leaf juice, about a cup, is drunk at morning to reduce body heat to treat sun-stroke and also for acidity.

(*iv*) Leaf juice is mixed in milk. A cup of it is consumed for two days twice a day to treat acidity.

(*v*) Fruit wall (epicarp) is levigated in water. This paste is applied on abdomen externally to cure intestinal torsion.

129. *Lycopersicon esculentum* Mill. (Solanaceae) Tamata

Medicine

(*i*) Consumption of ripe fruit is advised daily to increase red blood corpuscles.

130. *Madhuca longifolia* (Koen.) Macbr. (Sapotaceae) Mahu

Medicine

(*i*) Seeds after removal of seed coats are wrapped in leaves of betel (*Piper betle* L.) and then consumed to treat tonsils and throat swellings till cure.

(*ii*) Seeds are dried and powdered. It is dissolved in water and boiled. Two to three drops of this are poured into nose at night for seven days to treat fever in head.

(*iii*) Fruit juice is mixed with leaf juice of *Ipomoea carnea* Jacq. sp. *fistulosa* (Mart. and Choisy) Austin. Two spoonful of it is administered at morning for seven days to cure piles.

Miscellaneous Uses

(*i*) Flowers are used to prepare wine.

131. *Maerua oblongifolia* (Forsk.) A. Rich. (Capparidaceae)

Miscellaneous Uses

(*i*) Plants are used as fodder especially for goats.

132. *Mangifera indica* L. (Anacardiaceae) Amba

Medicine

(*i*) Kernels are baked and then consumed to control diarrhoea. Three to four kernels are advised for a day till cure.

(*ii*) Kernels are boiled in salt water. It is consumed to improve digestive capacity.

(*iii*) Latex after cutting leaves is applied on pimples.

(*iv*) Kernels are rubbed with the fruit of *Emblica officinalis* Gaertn. in equal amount with water and made into paste. This paste is applied on skull for growth of hair on bald.

Miscellaneous Uses

(*i*) Leaves are used to prepare festoons to decorate houses.

(*ii*) Wood is useful for fuel.

133. *Martynia annua* L. (Martyniaceae) Wagh-nakhi

Medicine

(*i*) Leaf juice is applied on stomach to kill the intestinal worms.

134. *Maytenus emarginata* (Willd.) Ding Hou (Celastraceae) Henkal

Medicine

(*i*) Young leaves are chewed to clear the mouth.

135. *Mentha spicata* L. (Lamiaceae) Pudina

Medicine

(*i*) Leaves are consumed to control dysentery. It is also useful for to reduce nausea.

(*ii*) Leaf juice is mixed with rhizome extract of *Zingiber officinale* Rosc. in 1:1 ratio. One spoonful of it is drunk for few days to cure cough.

(*iii*) The leaf juice is homogenized with honey. A spoonful of it is used to stop vomiting.

(*iv*) Leaf juice is mixed in curd. A cup of it is drunk daily before meal at morning for remove urinary complaints.

136. *Merremia aegyptia* (L.)Urb. (Convolvulaceae)

Miscellaneous Uses

(*i*) Leaves are used as fodder especially for goats and sheeps.

137. *Mimosa hamata* Willd. (Mimosaceae) Arti

Miscellaneous Uses

(*i*) Leaves are used as fodder for goats.

138. *Mimosa pudica* L. (Mimosaceae) Sharmi, Lajalu

Medicine

(*i*) Seeds and leaves are consumed along with the leaves of *Piper betle* L. as a remedy against mental complaints.

139. *Mitragyna parvifolia* (Roxb.) Korth. (Rubiaceae) Koham

Miscellaneous Uses

(*i*) Leaves are used as fodder for cattle.

(*ii*) Wood is used for house construction and agricultural implements.

140. *Momordica charantia* L. (Cucurbitaceae) Karle

Medicine

(*i*) Leaf juice, about a cup, is drunk to kill the intestinal worms.

(*ii*) Fruit juice is drunk daily. About a cup is advised for the patient suffering from diabetes. It is followed for few days.

(*iii*) One spoonful leaf juice mixed in a cup of cow's milk is drunk before meal to treat jaundice for three days.

(*iv*) Roots are rubbed in water. A spoonful of it is advised daily for the children to kill the intestinal worms till cure.

(*v*) Leaf juice is mixed with curd. A cup of it is drunk before meal for scant urination.

141. *Morinda pubescens* J. E. Sm. (Rubiaceae) Jangli-ramphal

Ethnoveterinary Medicine

(*i*) Leaves are fed to animals suffering from colic.

152. *Moringa oleifera* Lamk. (Moringaceae) Shevga

Medicine

(*i*) Leaves and fruits are cooked and consumed with bread by the patient suffering from jaundice. It is advised till cure.

Miscellaneous Uses

(*i*) Leaves and green pods are used as fodder for goats.

143. *Mucuna pruriens* (L.) DC. (Fabaceae) Kach-kuiri

Medicine

(*i*) 250 gm of seeds are boiled in a litre of milk. Seeds are soaked in water overnight. Seed-coats are then removed and dried in shadow. These are powdered. Two gm of this powder in a cup of milk is consumed for a month at morning to cure impotency.

(*ii*) Fruit spines are mixed in jaggery (*Saccharum officinarum* L.) and homogenized. Pellets are prepared. Two pellets are advised for seven days, before meal to kill intestinal worms.

144. *Mukia maderaspatana* L. (Cucurbitaceae) Kharwad

Miscellaneous Uses

(*i*) Plants are used as fodder for cattle.

(*ii*) Ripe berries are cherished by the local boys.

145. *Musa paradisiaca* L. (Musaceae) Keli

Ethnoveterinary Medicine

(*i*) Unripe fruit walls are made into paste and applied on wounds to remove worms from injury.

Miscellaneous Uses

(*i*) Plants are used to decorative during festivals and religious ceremonies.

(*ii*) Leaves are used to serve meal.

(*iii*) Inflorescence is used in worships and several festivals.

146. *Nerium indicum* Mill. (Apocynaceae) Kanher

Medicine

(*i*) Leaves are burnt and ash is mixed with coconut oil (*Cocos nucifera* L.). This paste is applied on burns.

147. *Nicotiana tabacum* L. (Solanaceae) Tambakhu

Medicine

(*i*) Oil of castor seeds (*Ricinus communis* L.) and cow's urine is applied on leaf. This leaf is tied in case of hornea.

Ethnoveterinary Medicine

(*i*) Leaves are mixed with sugar and camphor (*Ocimum kilimandscharium* Guercke.). This paste is applied on injury to kill worms.

148. *Nyctanthes arbor-tristis* L. (Oleaceae) Parijat

Medicine

(*i*) Leaves of this plant and those of *Vitex negundo* L. are stored in red bottle for a day in sunlight. It is then boiled. This decoction 25 ml twice a day is drunk to cure rheumatism till cure.

149. *Nymphaea nouchali* Burm. *f.* (Nymphaeaceae) Kamal

Medicine

(*i*) Fine powder of flowers is mixed with Sandal wood powder (*Santalum album* L.) and camphor (*Ocimum kilimandscharicum* Guercke). Its paste is applied on forehead for migraine for three to four days.

150. *Ocimum tenuiflorum* L. (Lamiaceae) Tulsi

Medicine

(*i*) Inflorescence is dried in shadow. It is powdered and homogenized with honey. A spoonful of it is drunk thrice a day to cure migraine.

(*ii*) Ten gm of root powder with ten gm of old (black) jaggery (*Saccharum officinarum* L.) is consumed to cure impotency. This is advised for a month twice a daily.

(*iii*) Leaf powder and powder of cloves (*Syzygium aromaticum* Merr. and Perry.) are mixed. It is added in teeth cavities to cure tooth-ache.

151. *Opuntia elatior* Mill. (Cactaceae) Fanta, Niwdung

Medicine

(*i*) Fruits are burnt and juice is extracted from them. This juice, about ten ml, is drunk to treat cough for three days daily.

(*ii*) Leaves are warmed and tied on injury or swelling.

(*iii*) Leaf juice is approved to cure scabies and eczema.

(*iv*) Slit is carved on the leaf. Turmeric powder (*Curcuma longa* L.) and salt is added in the slit. It is warmed and then rubbed on knees to cure knee-ache.

(*v*) Fruits are consumed to increase blood during anemic conditions.

152. *Oxalis corniculata* L. (Oxalidaceae) Ambushi

Miscellaneous Uses

(*i*) Leaves are cherished to allay thirst being sour.

(*ii*) Plant is used as fodder for cows, buffaloes and goats.

153. *Pentanema indicum* (L.) Ling. (Asteraceae)

Miscellaneous Uses

(*i*) Plant body is used as fodder for goats.

154. *Pergularia daemia* (Forsk.) Chiov. (Asclepiadaceae) Utran

Medicine

(*i*) Leaves are wrapped around joints daily to cure rheumatism till cure.

(*ii*) Roots are cooked in rice. It is consumed for seven days at morning to cure jaundice.

Ethnoveterinary Medicine

(*i*) Roots are fed to pet animals suffering from flatulence. These are mixed in their fodder.

155. *Phyllanthus airy-shawii* Burnel and Boux (Euphorbiaceae)

Miscellaneous Uses

(*i*) Plants are used as fodder for cattle.

156. *Phyllanthus amarus* Schumach. and Thonn. (Euphorbiaceae) Bhuiawala, Lekur-wali, Futane

Medicine

(*i*) The whole plants are boiled in water. A cup of decoction is given to a patient suffering from jaundice daily till cure.

157. *Phyllanthus maderaspatensis* L. (Euphorbiaceae) Bhui-awala

Miscellaneous Uses

(*i*) Plants are used as fodder for cattle.

158. *Physalis minima* L. (Solanaceae) Dokephodi

Medicine

(*i*) Leaves are powdered. A spoonful of powder is mixed with a cup of curd. It is consumed for seven days twice a day to cure piles.

159. *Piper nigrum* L. (Piperaceae) Mire

Medicine

(*i*) Seeds of this plant, flowering head of *Spilanthes calva* DC. (Akkalkadha) and Glycyrrhiza glabra L. (Jeshthamadh) and candy are powdered and mixed thoroughly. One spoonful is consumed daily for seven to eight days to clear voice.

(*ii*) Seeds are rubbed in water. This paste is applied on pimples.

160. *Pithecellobium dulce* (Roxb.) Benth. (Mimosaceae) Engraji-:chinch, Vilayati-chinch

Medicine

(*i*) Leaf juice, about two to three drops is dropped into ears to stop pus formation.

Miscellaneous Uses

(*i*) Leaves and green pods used as fodder for cattle and goats.

161. *Portulaca oleracea* L. (Portulacaceae) Ghol

Miscellaneous Uses

(*i*) Leaves are used as fodder for cattle.

(*ii*) Leaves are also used to prepare vegetable.

162. *Prosopis cineraria* (L.) Druce (Mimosaceae) Saundal

Medicine

(*i*) Leaves are crushed and the paste is applied on forehead for clear vision till cure.

163. *Psidium guajava* L. (Myrtaceae) Jam

Medicine

(*i*) Leaves are boiled in water. This decoction is used to gargle to treat mouth ulcer and for strong gums.

Miscellaneous Uses

(*i*) Wood is a good fuel.

164. *Punica granatum* L. (Punicaceae) Dalimb

Medicine

(*i*) Decoction of leaves is gargled or leaves are chewed to cure mouth ulcer.

(*ii*) Twenty one pieces of roots are tied on neck on Saturday for twenty one days to cure tooth-ache.

165. *Pupalia lappacea* (L.) Juss. (Amaranthaceae)

Miscellaneous Uses

(*i*) Plants are used as fodder for cattle.

166. *Raphanus sativus* L. (Brassicaceae) Mula

Medicine

(*i*) Curry of leaves is advised for the patient suffering from jaundice till cure.

(*ii*) Root slices are fried in cow's ghee. It is consumed at night for piles till cure.

167. *Rhus sinuata* Thunb. (Anacardiaceae) Amoni

Miscellaneous Uses

(*i*) Leaves are used as fodder for cattle.

(*ii*) Fruits are edible, especially unripe fruit cherished by local people.

168. *Rhynchosia minima* (L.) DC. (Fabaceae)

Miscellaneous Uses

(*i*) Plants are used as fodder for cattle and goats.

169. *Ricinus communis* L. (Euphorbiaceae) Erandi

Medicine

(*i*) Leaves are rubbed in cow's milk. One cup of it for three days at morning is given to the patient suffering from jaundice.

(*ii*) Few seeds are rubbed and consumed at morning without mouth cleaning for better stomach-complaints.

(*iii*) Roots are rubbed in cow's milk. A cup of it is administered for three days, twice a day, to treat colic.

(*iv*) Leaf juice, one spoon, is drunk and applied on abdomen to kill the intestinal worms in case of children.

(*v*) Seed oil, about 15 to 20 ml twice a day, is administered to check amoebic dysentery till cure.

(*vi*) Leaves are warmed after smearing coconut oil (*Cocos nucifera* L.). These are wrapped around stomach to treat flatulence.

170. *Rivea hypocrateriformis* (Desr.) Choisy. (Convolvulaceae) Fang

Miscellaneous Uses

(*i*) Leaves are used as fodder for all type of cattle.

(*ii*) Leaves are used as vegetable.

171. *Saccharum officinarum* L. (Poaceae) Oos

Medicine

(*i*) Fresh culms are squeezed for juice. The juice, about a glass, is advised for the patient suffering from jaundice.

172. *Salvadora persica* L. (Salvadoraceae) Pilu-kathar

Medicine

(*i*) Leaves are cooked and the knees are warmed with these to treat arthritis daily till cure.

173. *Santalum album* L. (Santalaceae) Chandan

Miscellaneous Uses

(*i*) Leaves are used as fodder for goat.

174. *Sapindus laurifolius* Vahl (Sapindaceae) Ritha

Medicine

(*i*) Fruit are burnt and their epicarps are consumed along with jaggery (*Saccharum officinarum* L.) for three days to cure piles.

Miscellaneous Uses

(*i*) Fruit is used to polish the jewellary.

(*ii*) Fruit are employed for hair washers.

175. *Schleichera oleosa* (Lour.) Oken. (Sapindaceae) Kusum

Miscellaneous Uses

(*i*) Leaves are used as fodder for cattle.

176. *Sclerocarpus africanus* Jacq. (Asteraceae)

Miscellaneous Uses

(*i*) Plant is used as fodder for cows and buffaloes.

177. *Semecarpus anacardium* L. *f.* (Anacardiaceae) Bibba

Medicine

(*i*) Two to three drops of seed oil are mixed with animal ghee. It is consumed thrice a day for three days to cure amoebic dysentery.

(*ii*) Paste of seed is applied on hill to cure hill-ache.

(*iii*) Five to six seeds are powdered and homogenized with 50 gm alum. It is used as tooth-paste daily to cure tooth-ache.

(*iv*) Seeds oil is mixed in a glass of unboiled milk. It is drunk at night before sleep to cure chronic cough and whooping cough as well.

Ethnoveterinary Medicine

(*i*) Seeds oil is applied on tongue of pet animals to treat mouth ulcer of animals.

(*ii*) Seed is homogenized with camphor (*Ocimum kilimandscharium* Guercke.). It is applied in eyes in sunlight to treat eye complaints.

178. *Sesamum orientale* L. (Pedaliaceae) Til

Medicine

(*i*) Stem powder is spread on injury caused due to axe.

179. *Sesbania sesban* (L.) Merr. (Fabaceae) Shewari

Miscellaneous Uses

(*i*) Leaves are used as fodder for goats and sheeps.

(*ii*) Woos is used as fuel.

(*iii*) These are planted usually to protect crops from winds.

180. *Sida acuta* Burm. *f.* (Malvaceae)

Miscellaneous Uses

(*i*) Leaves are fed to goats.

181. *Solanum anguivi* Lamk. (Solanaceae) Ubhi-ringani

Medicine

(*i*) Seeds and roots and honey are homogenized. It is consumed daily for thirty days to cure uterine problems.

182. *Solanum virginianum* L. (Solanaceae) Ringni

Medicine

(*i*) Roots are extracted in whey. About a glass is drunk thrice as antidote for snake-bite, especially for 'Ghonas', a species of serpent.

(*ii*) Seeds powder is mixed with honey in ratio 1:2. This mixture about one spoonful, is taken orally for cough till cure.

(*iii*) Seed powder is homogenized with honey. One spoonful for seven days is consumed by the patient suffering from asthma.

(*iv*) Fruit are dried in shadow and powdered. It is applied on gums for ten minutes at morning. This is used to cure toothache.

(*v*) Roots are levigated into cow's urine. This paste is applied to cure nail bed abscess.

183. *Sonchus oleraceus* L. (Asteraceae)

Miscellaneous Uses

(*i*) Leaves are used as fodder for cattle.

184. *Sorghum bicolor* L. (Poaceae) Chikni

Medicine

(*i*) Seeds are powdered and mixed in cow's urine. Pellets are prepared. Two pellets consumed twice a day at morning at night for twenty-one to treat leucoderma till cure.

185. *Spilanthes calva* DC. (Asteraceae) Akkalkadha

Medicine

(*i*) The children's are advised to chew fresh flowers as a remedy against stammering before meal.

(*ii*) Flowers are consumed at night to treat cough and to increase blood circulation.

(*iii*) Powder is made from the dried flowers, pepper seeds (*Piper nigrum* L.), cloves (*Syzygium aromaticum* Merr. and Perry.), cardamon seeds (*Elettaria cardamomum* L.), sugar (*Saccharum officinarum* L.). A cup of decoction prepared is used for fever till cure.

186. *Syzygium cumini* (L.) Skeels (Myrtaceae) Jambhul

Medicine

(*i*) Fruits are consumed to control the diabetes.

(*ii*) Fine powder of dried seeds (10-15gm) is given to children to control amoebic dysentery till cure.

(*iii*) Fine powdered is hold in mouth for a while to treat toothache.

(*iv*) Five to ten fruits are eaten before meal at morning to control various stomach complaints.

Miscellaneous Uses

(*i*) Wood is used to make rolling-pin.

(*ii*) Leaves are good fodder for cattle.

(*iii*) Twigs or leaves are placed on doors of abodes on the occasion of marriage and religious ceremony.

187. *Tagetes patula* L. (Asteraceae) Zendu

Medicine

(*i*) Receptacles of inflorescences are consumed to cure nausea and colic.

188. *Tamarindus indica* L. (esalpiniaceae) Chinch

Ethnoveterinary Medicine

(*i*) Leaves are cooked with ant-hill soil. This hot paste is applied on bone fracture or cramps.

189. *Tectona grandis* L. *f.* (Verbenaceae) Sag

Medicine

(*i*) Seeds are levigated in water and this paste is applied on stomach to treat scant urination.

Ethnoveterinary Medicine

(*i*) Seeds are levigated and fed to pet animals to treat stomach complaints.

Miscellaneous Uses

(*i*) Stem wood is used for making pestle and many other domestic and agricultural tools and implements, apart from house and hut construction.

190. *Terminalia arjuna* (Roxb.) Wight and Arn. (Combretaceae) Arjun-sadala

Medicine

(*i*) Bark is dried and powdered. This powder is applied on injury to avoid infection.

Ethnoveterinary Medicine

(*i*) Leaves are fed to the pet animals suffering from dysentery till cure.

191. *Terminalia crenulata* Roth (Combretaceae) Sandadi

Miscellaneous Uses

(*i*) Leaves are used for hut construction and preparing agricultural implements.

192. *Tinospora cordifolia* (Willd.) Miers. (Menispermaceae) Gulwel, Amrutwel

Medicine

(*i*) Dried stem powder is homogenized with honey. One spoonful of it is consumed per day to reduce acidity.

(*ii*) Stems are dried, powdered and mixed with candy. A spoonful of it is consumed twice daily to check body heat till cure.

(*iii*) Stems and leaves are dried and burnt. Ash is dissolved in water and boiled till salt is obtained. It is added in a spoonful water and drunk, to treat irregular menstruation.

(*iv*) The above salt is added in a cup of milk and given daily to cure jaundice for three days.

(*v*) Pieces of stem are boiled with rice grain (*Oryza sativa* L.). It is consumed for three days to cure jaundice.

(*vi*) Stem decoction, about a cup, is drunk for seven days to cure fever.

(*vii*) Stem and roots of *Cissampelaos pareira* L. are dried and powdered. A spoonful of it is consumed at morning before meal for twenty-one days to cure anaemia.

(*viii*) Take a root of this plant growing especially on *Azadirachta indica* A. Juss. Its decoction, about 15 ml, is advised to ladies for twenty one days for increasing milk.

(*ix*) Stems are powdered with roots of *Withania somnifera* (L.) Dunal. A spoonful of it is consumed with milk to treat impotency. It is advised for three weeks.

Miscellaneous Uses

(*i*) Stem is used to tie the bundles of fodder and fuel by the local people.

193. *Triticum aestivum* L. (Poaceae) Gahu

Medicine

(*i*) Juice (25ml) or few leaves of young seedlings consumed before meal to treat constipation and also for tooth-ache.

194. *Triumfetta annua* L. (Tiliaceae) Rodge

Miscellaneous Uses

(*i*) Leaves are used as fodder for cattle.

195. *Vigna mungo* (L.) Hepper (Fabaceae) Udid

Medicine

(*i*) Leaf paste is applied on injuries caused due to thorns and sickle.

196. *Vitex negundo* L. (Verbenaceae) Nirgudi

Medicine

(*i*) Leaves are boiled in water. This water is used by the patient suffering from rheumatism.

(*ii*) Brick is heated. Leaves are kept tightly on it. This brick is used to warm body to cure arthritis.

(*iii*) Leaf paste is used as tooth-paste at morning to cure tooth-ache.

Miscellaneous Uses

(*i*) Stems are employed to make grain containers.

197. *Withania somnifera* (L.) Dunal (Solanaceae) Ashwagandha, Askand

Medicine

(*i*) Roots are dried and powdered. A spoonful of it is homogenized in honey or milk. It is taken orally for twenty days to treat impotency.

(*ii*) Roots are dried and powdered. Two spoonful powder in a cup of cow's milk are added. It is administered for five days at morning to avoid 'night-dream' i.e. nocturnal emission and also for twenty-one days for body strength.

(*iii*) Root are dried in shadow and powdered. A spoonful powder in a cup of milk is consumed daily morning for seven days to cure heap-ache.

(*iv*) Root powder is mixed in a cup of cow's milk. Candy is added in it. This is consumed for twenty one days, twice a day to cure leucorrhoea.

(*v*) Root powder of rhizome of *Curcuma amada* Roxb. are homogenized in oil. Few drops are dropped into ears to stop pus formation.

198. *Xanthium indicum* L. (Asteraceae) Kokan, Gokru

Medicine

(*i*) Leaf juice is applied on bleeding injury due to thorns.

(*ii*) Leaf juice is applied on gums daily to cure teeth-ache.

199. *Zingiber officinale* Rosc. (Zingiberaceae) Aale, Sunth

Medicine

(*i*) Dried and treated rhizomes are called 'sunth'. A spoonful powder is mixed in a glass of milk. It is administered to a patient suffering from amoebic dysentery. This is advised at night till cure.

200. *Ziziphus mauritiana* Lamk. (Rhamnaceae) Ber

Medicine

(*i*) Two spoonful of leaf juice is drunk twice a day for a weak by a patient suffering from abdominal complaints.

(*ii*) Leaf juice is used as hair tonic to control hair loss. It is also applied on swelling.

(*iii*) Leaf paste is used to cure injury.

SECTION–III
(Chikhali, Mehkar and Buldhana Tehsils)

1. *Abrus precatorius* L. (Fabaceae) Lal-gunj

Medicine

(*i*) Few root pieces and some leaves are consumed for 2-3 days to clear throat and voice.

2. *Abutilon indicum* (L.) Sweet (Malvaceae) Dabba-zad

Ethnoveterinary Medicine

(*ii*) Leaves are fed to sheeps to treat against 'Lal-khurgati' i.e. FMD (Foot and Mouth Disease).

Miscellaneous Uses

(*ii*) It is used as fodder for cattle.

(*iii*) Stems are used for making partition wall of huts.

3. *Acacia chundra* Willd. (Mimosaceae) Khair

Medicine

(*i*) Bark and leaves are dried, crushed and powdered. A spoonful of it is dissolved in water which is then drunk as tonic. It is advised till the patient overcomes weakness.

4. *Acacia leucophloea* (Roxb.) Willd. (Mimosaceae) Hiwwar

Medicine

(*ii*) Leaves and fresh bark are extracted together. It is applied twice a day to cure sorosis till cure.

(*iii*) The leaves are powdered finely. Take one spoon is daily with honey or milk is consumed as a tonic.

Miscellaneous Uses

(*i*) Wood used as fuel.

(*ii*) Wood is used to make peg for pet animals.

(*iii*) Leaves and green pods are used as fodder for goats.

5. *Acacia nilotica* (L.) Willd. (Mimosaceae) Babhul

Medicine

(*i*) Leaves of this species and also of *Ziziphus mauritiana* Lamk. are extracted. It is boiled and reduced to a paste. This paste is applied to eyes as 'kajal' to cure bloating.

(*ii*) Gum is mixed in sweet-balls as a tonic.

(*iii*) The stem pieces are employed to clean teeth at morning.

Miscellaneous Uses

(*i*) The plant leaves and pods are used as fodder for sheeps and goats.

(*ii*) Wood is used for making various furniture, fixtures, bullocks-carts, hut and house construction.

6. *Acalypha malabarica* Muell.-Arg. (Euphorbiaceae)

Miscellaneous Uses

(*i*) It is used as fodder for cattle.

7. *Achyranthes aspera* L. (Amaranthaceae) Agheda

Medicine

(*i*) Leaf paste is mixed with jaggery. A pellet is prepared. Two pellets a day after tea at morning and at midday are administered to cure body pain and also to remove thorns from body.

Miscellaneous Uses

(*i*) The plant is used as a fodder.

(*ii*) It is used as fodder for goats before flowering.

8. *Aegle marmelos* (L.) Corr. (Rutaceae) Bel

Medicine

(*i*) Leaves (21) are kept in water overnight. This decoction, 3-4 spoonfuls, is drunk at morning to regularize blood-pressure.

(*ii*) Pulp of fruit is dried and powdered. One spoonful daily for eight days is consumed to cure physical weakness.

Ethnoveterinary Medicine

(*i*) The leaves are washed and heated slightly. Nine leaves are mixed with bread. It is consumed thrice a day to deworm wounds.

Miscellaneous Uses

(*i*) The leaves are used for worship of Lord Shiva.

9. *Ageratum conyzoides* L. (Asteraceae)

Miscellaneous Uses

(*i*) It is used as fodder for cattle.

10. *Aerva lanata* (L.) Juss. (Amaranthaceae) Pandhari-fuli

Medicine

() Leaf extract about a cup in 1:2 ratios with whey is drunk daily for 15 days to cure kidney stone. Tea is to be avoided during the period of treatment.

11. *Ailanthus excelsa* Roxb. (Simaroubaceae) Mahruk

Medicine

(*i*) Leaf juice is applied daily on body to treat leucoderma. It is applied till cure.

Miscellaneous Uses

(*ii*) Wood is used for construction of huts.

12. *Allium cepa* L. (Liliaceae) Kanda

Medicine

(*i*) Bulb is splited and quickly smelled to a unconscious person.

(*ii*) Bulb splits are also tied on ears during journey in hot months.

13. *Allium sativum* L. (Liliaceae) Lasun

Medicine

(*i*) Few bulblets are fried in oil and consumed daily for eight days to arouse sex.

(*ii*) Bulblets are tide around neck to cure cough till cure.

14. *Aloe vera* L. (Liliaceae) Korphad

Medicine

(*i*) Parts of leaf of this species with the leaf of *Piper betle* L. are heated on a pan. These are cut into small pieces. These are consumed thrice a day to cure whooping cough.

(*ii*) Leaf pulp, about 5-6 gm, is consumed once daily to cure diabetes till cure.

(*iii*) Leaf pulp is applied to cure burns till cure.

15. *Alysicarpus longifolius* Wight and Arn. (Fabaceae)

Miscellaneous Uses

(*i*) It is used as fodder for cattle.

16. *Amaranthus blitum* L. (Amaranthaceae) Poka, Pokla

Ethnoveterinary Medicine

(*i*) Entire plants are fed to milking sheep to increase milk production.

17. *Amaranthus polygonoides* L. (Amaranthaceae)

Miscellaneous Uses

(*i*) It is used as fodder for cattle.

18. *Amaranthus spinosus* L. (Amaranthaceae) Kathyamath

Medicine

(*i*) Leaves are cooked without oil and consumed with 'Jawar' (*Sorghum bicolor*) bread for 11-15 days to regain weight-loss.

Ethnoveterinary Medicine

(*i*) Extract of whole plant, about a liter, is administered to cows or buffaloes, bullocks etc. with oil to cure stomach-ache.

19. *Ampelocissus latifolia* (Roxb.) Planch. (Vitaceae) Ran-draksha

Ethnoveterinary Medicine

(*i*) The roots are crushed and powdered finely. It is mixed in a glass of water and fed to pet animal to cure flatulence or stomach-ache.

20. *Anisomeles indica* (L.) O. Kuntze (Lamiaceae) Uparsuri

Medicine

(*i*) Roots are crushed and then used to prepare decoction, a cup of this is drunk once daily as antipyretic till cure.

21. *Anogeissus latifolia* (Roxb.) ex DC. (Combretaceae) Dhamoda

Medicine

(*i*) Dried fruits and its gum are powdered. This powder is mixed while preparing sweet-balls. One sweet ball per day is consumed during post-delivery period to avoid infections or diseases.

Miscellaneous Uses

(*i*) Wood is valuable as fuel-wood.

22. *Annona squamosa* L. (Annonaceae) Sitaphal

Miscellaneous Uses

(*i*) Crushed leaves are used to prepare decoction. Two spoons of it are taken orally with some sugar for three days as anthelmintic.

(*ii*) Five to six leaves are extracted. Extract, 4-5 spoons is administered to a drunkard to reduce the effect of wine.

23. *Apluda mutica* L. (Poaceae) Shipi, Tikhadi

Medicine

(*i*) Leaf paste is prepared on hand-palm and applied on wounds. It is also applied for bites snake and scorpion sting.

(*ii*) Fresh leaves are dipped in water for sometime, and then this water is boiled and distilled. The oil thus obtained is massaged over joints daily for rheumatism till cure.

Miscellaneous Uses

(*i*) It is used as fodder for cattle.

24. *Argemone mexicana* L. (Papavaraceae) Kate-vilayati, Ringni

Medicine

(*i*) Root paste is applied as antidote on scorpion-sting.

(*ii*) Roots are strung into a necklace and kept around neck to cure jaundice till cure.

25. *Aristolochia bracteolata* Lamk. (Aristolochiaceae) Gindiyan, Gindhani

Medicine

(*i*) Leaves (4-5) of this species are consumed with the leaves of *Piper betle* L. for four days to cure stomach-ache till cure.

(*Ii*) The leaves are boiled in water. It is used to bath infants for good health.

26. *Asparagus racemosus* Willd. (Liliaceae) Shatavari

Medicine

(*i*) Entire plants are dried and powdered. This powder, about 5 gm, is consumed before tea. It benefits to overcome general body weakness and also helps to improve memory. It is advisable for 15-20 days at a stretch.

27. *Aspidoptyris cordata* (Heyne ex Wall). (Malpighiaceae) Jalki

Medicine

(*v*) Roots are dried and powdered finely. A spoonful of it once a day for three days is consumed in milk to check acidity.

28. *Azadirahcta indica* L. (Meliaceae) Neem

Medicine

(*i*) A cup of leaf extract twice a day is drunk to control diabetes till cure.

(*ii*) Leaves, green or dried ones, are kept in the grain containers to avoid insects.

(*iii*) Decoction of inner bark, about a cup, is drunk for three days to check stomach-ache.

(*iv*) Green fruit paste is directly applied on heads to reduce dandruff formation. It is advised for 3-4 days.

(*v*) Ripe fruit pulp is applied in eyes to cure eyespots.

Miscellaneous Uses

(*i*) Wood is used for making furniture and for hut or house construction.

29. *Balanites aegyptiaca* (L.) Del. (Balanitaceae) Fetri, Hingenbet

Medicine

(*i*) Fresh stem is cut into pieces and these are rubbed to obtain paste. It is applied on scorpion-sting.

(*ii*) Fruits are boiled in water and then the fruit is consumed as antidote for scorpion-sting or snake-bite.

Ethnoveterinary Medicine

(*i*) Eight to ten seeds are wrapped in the leaves of *Piper betle* L. They are fed to pet animals for flatulence.

30. *Baliospermum raziana* Keshav. *et* Yog. (Euphorbiaceae) Dati

Medicine

(*i*) Leaf extract is applied directly to cure ringworm.

(*ii*) The stem pieces are strung around neck of infants during teething. The strung of stem should be placed around the neck by one's maternal uncle.

31. *Bambusa arundinacea* Willd. (Poaceae) Bamboo

Medicine

(*i*) Small pieces of inner part of growing stems are consumed per day as body tonic. They are also cooked as a vegetable.

Miscellaneous Uses

(*i*) The plant is used to make huts, partition walls, ladder and domestic artifacts.

32. *Barleria prionitis* L. (Acanthaceae) Kate-korati

Medicine

(*i*) Leaves are chewed to cure mouth ulcer. It is advised once a day but till cure.

33. *Basella alba* L. (Basellaceae) Vavding, Bhajyanche-wel

Medicine

(*i*) Decoction of seeds is prepared in water. About two spoons are advised once a day to cure stomach-ache.

(*ii*) Leaves are used as vegetable to treat rheumatism.

34. *Biophytum sensitivum* (L.) DC. (Oxalidaceae) Lajalu

Medicine

(*i*) The whole plants are crushed and decoction is prepared. One cup is drunk once daily to cure rheumatism till cure.

(*ii*) Two drops of leaf juice are added in eyes to cure conjunctivitis till cure.

35. *Blepharis repens* (Vahl) Roth (Acanthaceae) Hadsan

Medicine

(*i*) Plants are cut into pieces, dried and powdered. It is mixed in sweet-balls. These are consumed as body tonic.

36. *Blumea membrancea* DC. (Asteraceae) Pivali-fuli

Medicine

(*i*) Two to three leaves of *Piper betle* L. and leaves of this species are extracted. It is filtered using cotton cloth. Two to three drops are added in an ear of opposite side of aching tooth. This is advised till cure.

37. *Bombax ceiba* L. (Bombacaceae) Kathshevar

Ethnoveterinary Medicine

(*i*) Bark is powdered and mixed in a litre of water. It is filtered and then administered to calf, cow or bullock to control dysentery.

Miscellaneous Uses

(*i*) Wood is used as fuel.

38. *Boswellia serrata* Roxb.ex Coleb. (Burseracaceae) Salai

Medicine

(*i*) Latex is applied twice daily on injuries till cure.

Miscellaneous Uses

(*i*) Wood is used for making windows and biting drums.

39. *Brassica juncea* (L.) Czerm. and Coss. (Brassicaceae) Mohari

Medicine

(*i*) Seeds are powdered finely. A spoon of it is mixed in curd. It is drunk once in a day as anathelmantic.

(*ii*) Seed oil is massaged on body thrice a day against rheumatism.

(*iii*) Seed powder is mixed in honey. It is applied on body to treat cramps.

40. *Buchanania lazan* Spreng. (Anacardiaceae) Char

Miscellaneous Uses

(*i*) Leaves are fed to cattle and also for making roofs.

(*ii*) Wood is used for hut construction.

41. *Butea monosperma* (Lamk.) Taub. (Fabaceae) Palas

Medicine

(*i*) Leaves are heated on a pan and then powdered. It is mixed in coconut oil and massaged twice a day on knees to treat rheumatism till cure.

(*ii*) Fruit walls are crushed and soaked in milk. Paste is prepared. It is applied on face and then washed after sometime. It is beneficial for pimples and to remove facial black spots.

(*iii*) The flower extract about one spoonful per day is drunk to improve urination.

Miscellaneous Uses

(*i*) Wood is used for making bullock-carts and furniture.

(*ii*) The leaves are used for preparing dining plates and bowls.

42. *Cajanus cajan* (L.) Millsp. (Fabaceae) Tur

Medicine

(*i*) Leaves are squeezed and the juice is applied twice a day to cure wounds.

Miscellaneous Uses

(*i*) Stems are used to make partition walls and roof of huts.

(*ii*) The stems are used to make sweeping brooms.

(*iii*) It is a good fuel.

43. *Calotropis gigantea* (L.) R. Br. (Asclepiadaceae) Pandhari-rui

Medicine

(*i*) About 5 to 6 leaves are roasted and then extracted. Two to three spoonful of extract mixed with milk is administered orally for few days to cure joint-ache.

(*ii*) Leaves are heated on a pan and wrapped on legs or hands to cure their stiffness.

(*iii*) Middle parts of flowers are taken out. Few flowers are consumed to cure cough and asthma.

(*iv*) Middle parts of few flowers along with few Cardamom are consumed for 5-7 days to cure piles.

44. *Calotropis procera* (Ait.) R. Br. (Asclepiadaceae) Rui (Lal)

Medicine

(*i*) Leaves are heated on a pan. They are wrapped daily around infected part of legs till cure.

45. *Capparis grandis* L. (Capparidaceae) Pachara

Medicine

(*i*) Small pieces are strung into a wreath. It is kept around the neck of a person suffering from fever till cure.

46. *Capsicum annuum* L. (Solanaceae) Mirchi

Medicine

(*i*) Crushed fruit paste is applied against scorpion bite.

47. *Cardiospermum halicacabum* (Sapindaceae) Fatakadi

Ethnoveterinary Medicine

(*i*) About 250 gm of leaves are crushed. The extract is given orally to pet animals to cure stomach-ache.

48. *Carissa congesta* Wight (Apocynaceae) Karwand

Medicine

(*i*) Fruits are cherished but also help to recover acidity.

Miscellaneous Uses

(*i*) Leaves are fed to goats.

(*ii*) The twigs are used for farming, farmyards and cultivated fields.

49. *Carthamus tinctorius* L. (Asteraceae) Kardai

Medicine

(*i*) Leaves are used as vegetable to improve appetite.

50. *Cassia obtusifolia* L. (Caesalpiniaceae) Deo-tarota

Ethnoveterinary Medicine

(*i*) Leaves are fed to animals to cure 'lalkhurgati' *i.e.* FMD (Foot and mouth disease).

Miscellaneous Uses

(*i*) Plants are employed for preparing roof of huts.

51. *Cassia tora* L. (Caesalpiniaceae) Tarota

Medicine

(*i*) Root paste is applied as antidote for scorpion sting.

(*ii*) Leaves are used as vegetable to cure fever.

Miscellaneous Uses

(*i*) The plants are used for making partition walls of huts.

52. *Cassine alberns* (Retz.) Kosterm. (Celastraceae) Bhutkes

Medicine

(*i*) Root powder is used to prepare decoction. A glass of it is drunk as antidote for snake-bite. Leaf extract, few drops, are poured into nose against head-ache.

Ethnoveterinary Medicine

(*i*) Green leaves are fed to cows and buffaloes to increase in milk production.

53. *Catharanthus rosesus* (L.) G. Don (Apocynaceae) Sadaphuli

Medicine

(*i*) Fresh young leaves (4-5) are mixed in bread and are consumed daily to cure diabetes. This is continued for 8-10 days.

54. *Cayratia trifolia* (L.) Domin (Vitaceae) Tivhan

Medicine

(*i*) Fresh leaves are extracted. Two to three spoons of it are dropped in ears against tooth-ache. It is used for 3-4 days.

55. *Celastrus paniculata* Willd. (Celastraceae) Kalkamuni

Medicine

(*i*) Fruits are dried and powdered. Spoonful of it mixed in a cup of milk once daily is administered as a tonic for 15 days.

56. *Celosia argentea* L. (Amaranthaceae) Kurdu

Medicine

(*i*) Seeds (5 gm) are powdered finely. It is mixed in milk. It is drunk once a day to cure kidney-stone. It is followed for a week.

57. *Centella asiatica* (L.) Urb. (Apiaceae) Brahmi

Medicine

(*i*) Leaves are crushed and apply on head for eight days as hair-tonic.

58. *Chenopodium murale* L. (Chenopodiaceae)

Miscellaneous Uses

(*i*) Plants are used as fodder for cattle.

59. *Chloris virgata* (Poaceae)

Miscellaneous Uses

(*i*) It is used as fodder for cattle.

60. *Chlorophytum borivilianum* Santapau and Fernad. (Liliaceae) Safed-musali

Medicine

(*i*) Roots tubers are dried and powdered. A spoonful with milk is drunk before sleep to keep fit or as tonic.

61. *Cicer arientinum* L. (Fabceae) Harbhara

Medicine

(*i*) Seed flour is mixed with a pinch of turmeric powder. This paste rubbed on face for eight days to remove facial black spots.

62. *Cissampelos pareira* L. (Menispermaceae) Dhavla vasu, Pandhara vasu

Medicine

(*i*) Roots are cleaned and its slices are wrapped in a leaf of *Piper betle* L. These are consumed thrice a day to cure hepatitis. It is advised for 7-8 days.

(*ii*) The paste of roots is diluted in water. One spoonful of it is administered to cure cough till cure.

63. *Citrullus colocynthis* (L.) Schrad. (Cucurbitaceae) Kadu-indravan

Medicine

(*i*) Leaves are squeezed to obtain juice. A spoonful of it once a day is drunk to treat general body weakness especially during post-delivery period.

64. *Citrus aurantifolia* (Christm.) Sw. (Rutaceae) Nimbu

Medicine

(*i*) Fruit juice is taken orally with honey against acidity.

(*ii*) Fruit juice +1 spoonful honey once daily for 30 days is beneficial to gain weight loss.

(*iii*) A fruit and one gm salt are mixed. It is heated on a flame. It is then consumed to cure acidity.

65. *Citrus sinensis* L. (Rutaceae) Mosambi

Medicine

(*i*) Decoction is prepared using leaves. One or two spoonful of it is drunk at morning for two consecutive days as antipyretic.

Miscellaneous Uses

(*i*) Wood is used as fuel.

66. *Cleome gynandra* L. (Capparidaceae) Ghanfod

Ethnoveterinary Medicine

(*i*) Leaf decoction, about half litre, is administered to pet animals daily as antipyretic till cure.

67. *Cleome viscosa* L. (Capparidaceae) Tilwand, Tiloni

Medicine

(*i*) Leaf extract of 2-3 leaves is dropped into ears to treat toothache. This is followed for 2-3 days.

(*ii*) Few drops of leaf extract are added in an ear of opposite side of aching-tooth till cure.

(*iii*) Leaf extract two to three drops are dropped in nose for two days in winter and 8-9 drops in summer to cure migraine.

(*iv*) A metal coin is dipped in leaf extract. It is tightly bound on inflamed veins. This is advised till cure.

68. *Clerodendrum multiflorum* (Burm. *f.*). (Verbenaceae) Katahankal, Arni

Medicine

(*i*) Fresh leaves are chewed thrice a day to cure mouth-ulcer.

(*ii*) Leaves are boiled and then crushed. Two spoonful of this twice a day is administered to cure stomach complaints.

Miscellaneous Uses

(*i*) It is used as a fodder particularly for goats and sheeps.

(*ii*) The twigs are used for roofs and partition walls of hut.

69. *Clitoria biflora* L. (Fabaceae) Jangli-wal, Gokarn

Medicine

(*i*) Vegetable of pods is cherished by the local people once a day to cure piles till cure.

Ethnoveterinary Medicine

(*i*) Extract of the leaves, about 5 ml, is poured in ears twice for two days to control worm in ears.

Miscellaneous Uses

(*i*) Dry fruits are kept in houses to bring happiness and silence.

70. *Coccinia grandis* (L.) Voigt. (Cucurbitaceae) Tondle

Medicine

(*i*) Raw fruit (one or two) are consumed to cure mouth-ulcer.

71. *Cocculus hirsutus* (L.) (Menispermaceae) Vasan

Medicine

(*i*) Leaves are extracted in water. Two spoonful of it twice a day for three days is consumed as a remedy against bloating.

(*ii*) Root bark is first removed. Roots are then crushed and made into paste after cooking them. It is applied on tumor daily until it disappears.

72. *Cocos nucifera* (Arecaceae) Naral

Medicine

(*i*) Coir is burnt. The ash about a spoon is mixed in water. It is drunk to stop hiccup.

73. *Coffea arabica* L. (Rubiaceae)

Medicine

(*i*) Coffee powder is mixed in milk and grape juice (*Vitis vinifera* L.) in equal quantity. A cup of it is drunk daily for 15 days to treat low blood pressure.

(*ii*) Coffee powder is applied on injuries or wounds to stop bleeding.

74. *Coix gigantea* Roxb. (Poaceae)

Miscellaneous Uses

(*i*) It is used as fodder for crops.

75. *Colocasia esculenta* (L.) Schott. (Araceae) Alu

Medicine

(*i*) Fruit paste is applied on face to remove black spots. It is advised for three days.

76. *Commelina benghalensis* L. (Commelinaceae) Nil

Miscellaneous Uses

(*i*) It is used as fodder for cattle.

(*ii*) Leaves are used as vegetable.

77. *Commelina forsklaei* Vahl (Commelinaceae) Kena

Miscellaneous Uses

(*i*) It is used as fodder for cattle.

(*ii*) Leaves are used as vegetable.

78. *Corchorus trilocularis* L. (Tiliaceae)

Miscellaneous Uses

(*i*) It is used as fodder for cattle.

79. *Cordia dichotoma* (L.) Forst. *f.* (Boraginaceae) Bhokar

Medicine

(*i*) Bark powder is mixed in coconut oil. This paste is applied on burns till cure.

80. *Cordia gharaf* (Forsk.) Ehrenb. (Boraginaceae) Gondhan

Medicine

(*i*) Leaves and stem barks are extracted. Two drops at a time for three days are administered to cure eye injuries.

(*ii*) Fruits (20-25) are consumed daily to stop diarrhoea till cure.

Miscellaneous Uses

(*i*) Wood is used for furniture.

(*ii*) Wood is used as fuel.

(*iii*) Leaves are fed to goats, cows and bullocks.

81. *Cordia macleodii* (Griff.) Hook. *f.* (Boraginaceae) Dahipalas

Miscellaneous Uses

(*i*) Wood is used for making windows, doors, and bullock carts.

(*ii*) Ripe fruits are cherished by the local people.

82. *Coriandrum sativum* L. (Apiaceae) Kothimbir, sambhar

Medicine

(*i*) A spoonful of seed powder mixed in water twice a day is administered to cure acidity. It is advised till cure.

83. *Crotalaria filipes* Benth. (Fabaceae) Fatfati

Miscellaneous Uses

(*i*) The plant is used as fodder cattle.

84. *Crotalaria juncea* L. (Fabaceae) Tag

Medicine

(*i*) Seeds are powdered. It is mixed in water. A cup of it is administered daily for 7-8 days for abortion.

85. *Cucumis melo* L. var. *utilissimus* (Roxb.) Duthie & Fuller (Cucurbitaceae) Kakdi

Medicine

(*i*) Pieces of raw fruits are kept on eyes to cure eye pains or to reduce burning of eyes.

86. *Cucurbita moschata* (Dutch. ex Lamk.) Poir: (Cucurbitaceae) Kashiphal, Gangaphal

Medicine

(*i*) Fruit slices are placed on burns for cooling effect.

87. *Curcuma amada* Roxb. (Zingiberaceae) Aambi-halad

Medicine

(*i*) Rhizomes are dried and powdered with common salt. Pellets are prepared. One pellet a daily for a month is consumed to cure irregular menstruation.

(*ii*) Rhizomes are cooked and applied on inflamed.

88. *Curcuma longa* L. (Zingiberaceae) Haldi

Medicine

(*i*) Rhizomes are powdered finely. Its paste is applied to cure body wounds.

(*ii*) Rhizomes are baked and then powdered. It is licked to cure cough. It is practiced twice a day till cure.

Ethnoveterinary Medicine

(*i*) Root juice is mixed in human urine in 1:1 ratio. It is orally administered to an animal suffering from stomach-ache or flatulence. About 2-3 glass of this mixture is advised twice a day.

89. *Cuscuta chinensis* Lamk. (Cuscutaceae) Amarwel

Medicine

(*i*) The whole plant is crushed in sheep's urine. It is cooked to prepare a gel. It is applied on the body once for eight day against rheumatism.

(*ii*) Plant juice is applied on body to treat pimples. It is advised once for three days.

Ethnoveterinary Medicine

(*i*) Plants are crushed and fed to milking animals to increase milk yield.

90. *Cyanotis cristata* L. (Commelinaceae)

Miscellaneous Uses

(*i*) It is used as fodder for cattle.

91. *Cymbopogan martini* (Poaceae) Rohis

Medicine

(*i*) The seeds are crushed first. It is then filled in cigars and smoked as a remedy against scorpion sting.

Ethnoveterinary Medicine

(*i*) Leaves are boiled in water. This decoction, about a litre, is administered to the domestic animals to check diarrhoea till cure.

92. *Cynodon dactylon* Pers. (Poaceae) Harali

Medicine

(*i*) Entire plant is used to prepare decoction. Half cup of it is drunk once in a day to reduce body heat.

Miscellaneous Uses

(*i*) The leaves are offered to Lord Ganesha.

(*ii*) It is used as a fodder for cattle.

93. *Cyperus bulbosus* Vahl (Cyperaceae) Lavhale

Miscellaneous Use

(*i*) It is used as fodder for cattle.

94. *Cyperus nutans* Vahl (Cyperaceae) Lavhale

Miscellaneous Use

(*i*) It is used as fodder for cattle.

95. *Cyperus rotundus* L. (Cyperaceae) Lavali, Nagarmotha

Medicine

(*i*) Leaf extract, about one spoonful, is mixed in water or milk to control acidity. It is consumed for 3-4 days.

96. *Dactyloctenium aegyptium* (Poaceae)

Miscellaneous Use

(*i*) It is used as fodder for cattle.

97. *Datura metel* L. (Solanaceae) Kala-dhotra

Medicine

(*i*) Extract of leaves is applied on forehead to cure headache till cure.

(*ii*) Root are crushed and consumed for 2-3 days with bread against hepatitis.

98. ***Delonix regia*** **(Boj. ex Hook.) Raf. (Caesalpiniaceae) Balamkakdi, Gulmohar**

Medicine

(*iii*) Seeds are crushed in cow's urine. This paste is applied on thigh to cure cramps.

Miscellaneous Uses

(*i*) The wood is used as fuel.

99. ***Desmodium dichotomum*** **(Willd.) DC. (Fabaceae)**

Miscellaneous Uses

(*i*) It is used as fodder for cattle, goats, cows and buffaloes.

100. ***Digera muricata*** **(L.) Mart. (Amaranthaceae) Lahan-kirdu**

Medicine

(*i*) Leaves are cooked and consumed as vegetable once a day to cure mouth ulcer and to remove foul smell. This also helps to control diarrhoea and dysentery.

101. ***Dioscorea bulbifera*** **L. (Dioscoreaceae) Jata-shankar**

Miscellaneous Uses

(*i*) Seeds are crushed and cooked. It is then applied to cure wounds.

102. ***Diplocyclos palmatus*** **(L.) Jeffery (Cucurbitaceae) Shivlingi**

Medicine

(*i*) After removal of seeds, fruits are wrapped by the leaves of *Piper betle* L. (Piperaceae). Pellets are prepared from these. Two pellets a day for three days are advised to regulate menstrual cycle.

103. ***Dolichandrone falcata*** **(Wall. ex DC.) (Bignoniaceae) Medhsing**

Medicine

(*i*) Fresh leaves are dried. They are crushed and boiled. A cup of this decoction is drunk twice a day to cure weakness during post-delivery period.

(*ii*) Gum and powder are used to prepare sweet-balls. They are consumed as tonic.

104. *Drimia indica* (Roxb.) Jessop (Liliaceae) Jangali-kanda

Medicine

(*i*) Roots and its bulb are crushed into a paste. It is applied on body thrice a day to cure eczema.

105. *Eclipta prostrata* (L.) L. (Asteraceae) Kala-maka

Medicine

(*i*) Fresh leaves are crushed. The juice obtained is applied to cure wounds and white skin patches thrice a day till cure.

(*ii*) Leaf juice is mixed in coconut oil in 1:3 ratio. It is applied on hair for eight days as a hair tonic.

(*iii*) Leaf juice or paste is applied directly on forehead, twice a day to relieve from fever till cure.

(*iv*) Seed powder is mixed with water. Two spoon of it is administered to a patient suffering from diabetes for 7 days.

(*v*) Stem is tied around neck or stomach to protect from snake.

106. *Echinops echinatus* Roxb. (Asteraceae) Utkatya

Medicine

(*i*) Roots are crushed and powdered finely. One spoonful of it is mixed in water and drunk at morning for 21 days to cure fever or typhoid.

107. *Ehretia asperra* Roxb. (Ehretiaceae) Lokhandi

Medicine

(*i*) Stem bark, roots and leaves are burnt. The ash is used as tooth-powder. This helps to cure tooth-ache and to cure mouth-ulcer and to remove foul smell. It should be used for 3-4 days.

108. *Elettaria cardamomum* L. (Zingiberaceae) Veldoda

Medicine

(*i*) Consumed one fruit when indigestion is experienced after consumption of bananas.

(*ii*) Fruits are wrapped in leaves of *Piper betle* L. It is consumed before tea to cure rheumatism for 7-8 days.

109. *Elyonurus royalean* ex A. Rich. (Poaceae)

Miscellaneous Uses

(*i*) It is used as a fodder for cattle.

110. *Emblica officinalis* Gaertn. (Euphorbiaceae) Aawala

Medicine

(*i*) Fruits are boiled and mixed in honey to prepare edible 'awalkanthi'. One or two awalkanthi are consumed daily for better health.

111. *Enicostemna axillare* (Lamb.) Raynal (Gentianaceae) Nay

Medicine

(*i*) Entire plants are dried and powdered finely. One spoonful of it once for two days is consumed with honey as tonic.

(*ii*) Decoction of entire plant is prepared in water. A cup of it is once for eight days drunk to cure kidney-stone. It is advised till cure.

112. *Eucalyptus camaldulensis* Dehnh. (Myrtaceae) Nilgiri

Medicine

(*i*) Leaf juice is squeezed. It is diluted in water in 1:1 ratio. One or two spoons of it is administered to cure cough twice a day till cure.

Miscellaneous Uses

(*i*) Wood used to make pillars of hut.

113. *Euphorbia indica* Lamk. (Euphorbiaceae) Lal-dudhi

Ethnoveterinary Medicine

(*i*) Entire plants are fed to milching sheeps, cows and buffaloes to increase milk production for a fortnight.

(*ii*) Decoction of entire plants, about half litre, is administered to pet animals twice a day to cure FMD (Foot and Mouth Disease).

114. *Euphorbia neriifolia* L. (Euphorbiaceae) Sabar-kand

Medicine

(*i*) Stem is heated on a flame and then juice from it is obtained. A spoonful of it is drunk against cough. It is advised for three days.

115. *Euphorbia prunifolia* Jacq. (Euphorbiaceae)

Miscellaneous Uses

(*i*) It is used as fodder for goats.

116. *Ficus benghalensis* L. (Moraceae) Wad

Medicine

(*i*) Newly formed prop roots are extracted. The extract is mixed in coconut oil in 1:1 ratio. It is applied on head daily to stop hair fall till cure.

(*ii*) Latex is applied on 'Batasha', an article of sweetmeat. One Batasha is consumed daily for 15 days to arouse sex and to improve quality and quantity of sperms.

(*iii*) The prop roots are extracted. The extract about spoonful once a day is consumed to increase milk for 7 to 8 days.

(*ii*) Fresh stem is crushed and applied on tongue for 4-5 times a day to cure cough.

117. *Ficus racemosa* L. (Moraceae) Umber

Medicine

(*i*) Fruits are fried in ghee. They are consumed daily as brain tonic.

(*ii*) Ripe fruits are selected in odd numbers. They are dried and powdered. One spoonful thrice for 11 days is consumed to cure irregular menstruation.

(*iii*) Latex is applied in mouth for three to four days to cure mouth ulcer.

(*iv*) Roots and fruits are used to prepare decoction. A cup of it is drunk against impotency. It is followed a fortnight or so.

118. *Ficus religiosa* L. (Moraceae) Pimpal

Medicine

(*i*) Warmed leaves are placed on organ or body part to treat stiffness.

(*ii*) Bark powder is homogenized in ghee. This paste is applied on organ to treat stiffness of legs or hands.

(*iii*) Small pieces of roots are placed around neck of mentally retarded person.

(*iv*) Curd and rice mixed with leaf paste are fed to children against stammering.

(*v*) Two drops of leaf juice are poured in both ears to combat snake-bite.

119. *Flacourtia indica* (Burm.*f.*) Merr. (Flacourtiaceae) Aakhadi

Medicine

(*i*) Fruits are dried and powdered. A spoonful of it twice a day is consumed in honey for a week to cure physical weakness.

Miscellaneous Uses

(*i*) The twigs are used as partition wall of huts.

120. *Foeniculum vulgare* Mill. (Apiaceae) Badishop

Medicine

(*i*) Seeds are crushed in water. The extract, about half cup, is drunk as a remedy against acidity. It is advised for few days till cure.

121. *Gliricidia sepium* (Jacq.) Kunth ex Walp. (Fabaceae)

Miscellaneous Uses

(*i*) Bunch of flowers are kept in house corners to avoid entrance of rats.

122. *Grewia flavescens* A. Juss. (Tiliaceae) Khadak-chopdi

Medicine

(*i*) The roots are cut into pieces, dried and powdered finely. One teaspoon powder is mixed in a glass of whey. This is drunk for seven days against piles. Sweet and spicy food is to be avoided.

123. *Hardwickia binata* Roxb. (Caesalpiniaceae) Anjan

Miscellaneous Uses

(*i*) Wood is used as fuel.

(*ii*) Wood is used for making furniture.

124. *Helicteres isora* L. (Sterculiaceae) Murad-sheng

Medicine

(*i*) Pods are rubbed in water. A spoonful of this paste is administered at the interval of two to three hours to control dysentery.

125. *Heteropogon contortus* R.&S. (Poaceae) Nil

Miscellaneous Uses

(*i*) It is used as fodder for cattle before flowering.

126. *Hibiscus caesius* Garcke (Malvaceae) Ambadi

Medicine

(*i*) Leaf vegetable is consumed or decoction of leaves, about a cup twice a day is drunk, to cure acidity till cure.

Miscellaneous Uses

(*i*) The stems are used for making partition walls.

127. *Hibiscus micranthus* L. (Malvaceae) Jungli-jawas

Medicine

(*i*) The whole plants are crushed and boiled in water, boiled, till it becomes thick. This is taken orally about 15 ml, for three days to cure bloating.

128. *Hibiscus trionum* L. (Malvaceae) Kurmudi

Medicine

(*i*) Fresh leaves are crushed on hand palms to remove extract. It is applied around ears to cure tooth-ache.

129. *Holarrhenu pubescens* (Buch.-Ham) Wall.ex G. Don (Apocynaceae) Dudhakanda

Medicine

(*i*) Fruit juice is applied on inner side of pot containing milk to turn it into curd.

(*ii*) Green pods are consumed with bread during post-delivery period to increase milk flow.

Ethnoveterinary Medicine

(*i*) Pods are mixed in fodder of cows, sheep and buffaloes to increase milk for a fortnight.

Miscellaneous Uses

(*i*) The wood is used for making furniture and also a fuel.

130. *Hordeum vulgare* L. (Poaceae) Javas

Ethnoveterinary Medicine

(*i*) Seeds are cooked. Their paste is applied on tumours.

(*ii*) Seed oil is applied on tumour until it disappears.

131. *Ipomoea hederifolia* L. (Convolvulaceae) Gohda

Ethnoveterinary Medicine

(*i*) Leaf extract is applied on the head around horns to kill worms.

132. *Jatropha gossypifolia* L. (Euphorbiaceae) Chandrajyoti

Medicine

(*i*) Wash the teeth with stem to make them strong and cure tooth-ache.

(*ii*) Plant latex diluted in water is gargled for three days to remove foul smell of mouth.

133. *Justicia adhatoda* L. (Acanthaceae) Adulsa

Medicine

(*ii*) A decoction of leaves, about a two spoons twice for 2-3 days to cure cough.

134. *Kalanchoe pinnata* (Lamk.) Pers. (Crassulaceae) Panfuti

Medicine

(*i*) Leaf extract is applied around ears to cure ear-ache.

(*ii*) Leaf extract, sugar cane (*Saccharum officinarum* L.) and honey 1:1:1 ratio (one spoon each) is homogenized and consumed for seven days to treat kidney-stone.

135. *Kigelia africana* (Lamk.) Benth. (Bignoniaceae) Bhend-fal

Medicine

(*i*) Small pieces of fruits are cooked. Paste is prepared from it. It is applied on entire body to check fever.

(*ii*) The fruit is cooked. About 8-10 gm of it is advised for women for conception for 15 days.

136. *Kirganelia reticulata* (Poir.) Baill. (Euphorbiaceae) Karad-kamuni

Medicine

(*i*) Fruits and leaves are consumed daily with jaggery for three days against rheumatism.

(*ii*) Ripe fruits are crushed in edible oil and applied directly on body to treat bloating.

137. *Lagenaria siceraria* (Molina) Standl. (Cucurbitaceae) Bhopla

Medicine

(*i*) Raw fruit is extracted. A cup of it is drunk daily to increase weight of body. It is followed for 15-20 days.

(*ii*) Fruit extract+ equal quantity of water, two Tulsi leaves (*Ocimum tenuiflorum* L.). Few leaves of Pudina (*Mentha piperata* L.) and few Miris (seeds of *Piper nigrum*) are crushed.

(*ii*) Decoction of fruit is prepared. One cup per day is advised for a patient suffering from blood pressure, body weight loss or for heat problem.

138. *Lagerstroemia parviflora* Roxb. (Lythraceae) Landga, Bondara

Medicine

(*i*) Two to three leaves are crushed on hand palms. It is applied directly on wounds. It is advised till cure.

Ethnoveterinary Medicine

(*i*) Leaves of this species and leaves of *Tamarandus indica* L. are homogenized with heated soil of ant-hill. It is applied on bone fracture till cure.

Miscellaneous Uses

(*i*) The wood is useful as fuel.

139. *Launaea procumbens* (Roxb.) Ramayya and Rajgopal (Asteraceae) Pathri

Ethnoveterinary Medicine

(*i*) Leaves are used as fodder to milching animals.

140. *Lawsonia inermis* L. (Lythraceae) Mehandi

Miscellaneous Uses

(*i*) Leaves fresh or dried are powdered. Its paste is used for staining body parts.

141. *Lepidagathis cuspidata* Nees (Acanthaceae) Benda-ghas

Medicine

(*i*) Decoction of leaves, about half cup, is drunk once for three to four days to cure diphtheria.

142. *Leucaena latisiliqua* (L.) Gill. (Mimosaceae) Subabhul

Ethnoveterinary Medicine

(*i*) Leaves are extracted. It is given to sheep and goats as tonic for few days.

Miscellaneous Uses

(*i*) Leaves and green pods are used as fodder for all pet animals.

(*ii*) Wood is used as pillars for hut construction.

(*iii*) Wood is also used as fuel.

143. *Mangifera indica* L. (Anacardiaceae) Amba

Medicine

(*i*) Cotyledons are removed from the seed. They are rubbed on stone with water and applied on a scorpion-sting.

(*ii*) Bark is powdered finely and sieved through cloth. It is mixed in coconut oil. It is applied daily to cure wounds.

Miscellaneous Uses

(*i*) Wood is used for making furniture.

(*ii*) The leaves are strung into a 'Toran', a wreath to decorate the entrance or door at the time of festivals and worships.

144. *Maytenus emarginata* (Willd.) Ding Hou (Celestraceae) Yenkal

Medicine

(*i*) Tender leaves are chewed for 2-3 days as a remedy against mouth ulcer and tooth-ache.

Miscellaneous Uses

(*i*) Leaves are fed especially for goats.

(*ii*) The twigs are used for hedges around farming, farmyards, and cultivated fields.

145. *Mentha spicata* L. (Lamiaceae) Pudina

Medicine

(*i*) Leaf extract, about a spoon, diluted with water twice a day is drunk against acidity till cure.

(*ii*) Leaves are crushed with fruit powder of *Emblica officinalis* Gaertn. It is used for hair washing to remove dandruff.

146. *Merremia dissecta* (Jacq.) Hall. (Convolvulaceae) Shapi-mischar

Medicine

(*i*) Extract of dry seeds is prepared in water. It is administered orally to a person against poisoning. About half cup extract is used at a time. Two to three doses are advised.

(*ii*) Plants are growing near houses to avoid entry of snakes.

147. *Mimosa hamata* Willd. (Mimosaceae) Aarati

Medicine

(*i*) Extract of seedlings is mixed with equal quantity of milk and honey. A cup of this mixture is administered against snake-bite at a time. Three to four times, if necessary is also advised.

148. *Momordica charantia* L. (Cucurbitaceae) Karle

Medicine

(*i*) Raw fruit juice, about 10-15 ml, is drunk daily to cure diabetes till cure.

149. *Morinda pubescens* J.E. Sm. (Rubiaceae) Bardoi

Medicine

(*i*) Bark is dried and powdered. One spoonful of it in whey daily is drunk for few days as antiemetic.

150. *Moringa concanensis* Nimmo *ex* Dalz. (Moringaceae) Jungli-Shevaga

Medicine

(*i*) Bark is heated and powdered. It is mixed in coconut oil (*Cocos nucifera* L.) and applied daily on injury. It is advised till cure.

151. *Moringa oleifera* Lamk. (Moringaceae) God-Shevga

Medicine

(*i*) Fruits are consumed as vegetable for 10-15 days to increase sperm quantity.

(*ii*) Leaf decoction and also leaf paste is applied on joints daily to cure rheumatism till cure.

(*ii*) Stem bark powder is mixed in juice of its leaves. This paste is applied on joints for a week to treat rheumatism.

152. *Mucuna pruriens* (L.) DC. (Fabaceae) Kach-kuiri

Medicine

(*i*) Fruit hairs are removed, dried and powdered. One spoonful powder mixed in water is advised as tonic for 7-15 days.

(*ii*) Fruit hairs are removed, powdered and mixed in jaggery. Pellets are prepared. Two pellets per day are advised as antithelmentic.

153. *Murraya koenigii* (L.) Sperng. (Rutaceae) Godnim, Kadhi-patta

Medicine

(*i*) Red fruits (10-15) consumed twice a day for 8-10 days to increase sperm count.

(*ii*) Inner bark of stem is crushed and then decoction is prepared. A cup of it, twice a day, against viral fever is administered for 2-3 days.

154. *Musa paradisiaca* L. (Musaceae) Kela

Miscellaneous Uses

(*i*) Leaves are used as dining plates and bowls.

155. *Nerium indicum* Mill. (Apocynaceae) Kanher

Medicine

(*i*) Leaves are crushed and then decoction is prepared. One cup of it is administered before tea to cure tuberculosis. It is advised till cure.

(*ii*) Roots are made into paste. One spoonful once a day is consumed with honey to cure piles.

Ethnoveterinary Medicine

(*i*) Crushed leaves are used to obtain decoction. One glass of it is daily given to pet animals for 3-4 days to control fever.

156. *Nicotiana tabacum* L. (Solanaceae) Tambakhu

Medicine

(*i*) Leaves are dried and powdered finely. It is mixed in lime, camphor and naphthalene thoroughly. The wound is washed and then this mixture is applied twice a day till cure.

157. *Nymphaea pubescens* Willd. (Nymphaaceae) Kamal

Medicine

(*i*) Seeds are powdered finely. It is mixed with honey. Two to three spoonful of it is consumed daily as a remedy against paralysis till cure.

158. *Ocimum americanum* L. (Lamiaceae) Ran-tulas

Medicine

(*i*) Seeds are crushed and powdered finely. It is mixed in water. This water is applied to eyes to cure eye infection. It is applied twice a day till cure.

(*ii*) Extract of leaf is applied on body to repel mosquitoes.

159. *Ocimum basilicum* L. (Lamiaceae) Tulsi, Sabja

Medicine

(*i*) Leaf extract, about spoonful, diluted in water is drunk against acidity for 4-5 days.

(*ii*) Leaves are crushed and mixed in honey. Two spoons are advised daily for two days against cough.

(*iii*) Inflorescence is dried and powdered finely. One spoonful per day in water is drunk for conception or for treating irregular menstruation.

(*iv*) Leaf extract is prepared in honey. Two to three drops are added in eyes to treat cataract.

(*v*) Leaf juice two drops, is dropped in ears to cure ear-ache.

160. *Opuntia elatior* Mill. (Cactaceae) Nivdung

Medicine

(*i*) Fruits are burnt on flame. One fruit daily for two days are consumed to cure cough.

161. *Pennisetum purpureum* Schumach. (Poaceae) Bhadli

Medicine

(*i*) Seeds are powdered and sweet-balls are prepared. They are consumed one per day at morning to cure weakness during pregnancy.

Miscellaneous Uses

(*i*) The plant is useful as fodder.

162. *Pergularia daemia* (Forsk.) Chiov. (Asclepiadaceae) Fispadri, Utran

Medicine

(*i*) Root paste is applied on the wounds daily.

(*ii*) A spoonful of leaf juice is taken orally to cure cough daily for three days.

(*iii*) Leaf extract is heated till it becomes thick. It is then applied for bone fracture.

(*iv*) The leaves are boiled in water. It is used to bath to cure body cramp.

(*v*) The leaf decoction is drunk, two spoonful twice a day, to cure diphtheria till cure.

(*vi*) The leaves are roasted with the leaves of *Calotropis procera* L. Extract of these is taken orally once a day to stomach-ache in case of infants.

163. *Phyllanthus amarus* Schumach and Thonn. (Euphorbiaceae) Bhui-awla

Medicine

(*i*) Entire plant is dried and powdered. It is used as tooth-powder. It is applied twice a day for about a week to treat tooth-ache and bleeding gums.

164. *Physalis minima* L. (Solanaceae) Doke-fodi

Medicine

(*i*) Decoction of leaves is administered once a day to pet animals as a remedy against flatulence. Four to five glasses of decoction are advised for two days.

165. *Plumeria alba* L. (Apocynaceae) Chafa

Medicine

(*i*) Small pieces of flowers are deep in coconut oil. These are made into paste and used twice a day to cure wound, scabies and ring-worms.

166. *Polygonum glabrum* Willd. (Polygonaceae)

Miscellaneous Uses

(*i*) It is used as fodder for cattle.

167. *Pongamia pinnata* (L.) Pierre (Fabaceae) Karanj

Medicine

(*i*) Leaf extract is mixed in oil. It is applied on knees for knee-ache and rheumatism till cure.

(*ii*) Fruits are dried and powdered. One spoonful of it is daily consumed with milk or honey to treat rheumatism till cure.

168. *Portulaca oleracea* L. (Portulacaceae) Ghol

Medicine

(*i*) Leaves are crushed and juice is squeezed. Two drops of this juice twice a day is dropped into ears to reduce ear-pains.

169. *Prosopis cineraria* (L.) Druce (Mimosaceae) Saundhad

Medicine

(*i*) Rice grains are soaked in water for two days. It is then crushed and mixed with butter and leaf juice. It is heated gently and applied thrice a day on body to cure 'Nagin' disease of skin.

(*ii*) Fresh bark is extracted. It is applied twice a day to cure sorosis.

170. *Psidium guajava* L. (Myrtaceae) Peru

Medicine

(*i*) One fresh leaves with catechu powder is consumed once a day to cure mouth ulcer.

171. *Pterocarpus marsupium* Roxb. (Fabaceae) Ragat-bel

Medicine

(*i*) Stem bark is crushed. Decoction is prepared. A cup of it is administered twice a day for three days to cure asthma. It is also useful for blood circulation chest-pain and cough.

172. *Punica granatum* L. (Punicaceae) Dalimb

Medicine

(*i*) Leaf extract and a pinch of sugar are homogenized. About half cup is consumed against dehydration.

173. *Pupalia lappacea* L. (Amaranthaceae)

Miscellaneous Uses

(*i*) It is used as fodder for cattle.

174. *Ricinus communis* L. (Euphorbiaceae) Arandi

Medicine

(*i*) Seeds extract, about one spoon, once a day is drunk to stop dysentery and dehydration.

(*ii*) Crushed fruits along with seeds are extracted. It is mixed in milk. Two spoons of extract are mixed in milk and are administered twice a day to cure hepatitis till cure.

(*iii*) Leaf decoction is prepared. About half cup once a day is taken orally to treat hepatitis for 5-6 days.

(*iv*) One spoonful of seed oil is administered against flatulence.

175. *Rosa damascena* Mill. (Rosaceae) Gulab

Medicine

(*i*) Crushed corollas are made into paste and applied for smoothness of face. Every time fresh corollas are used. Their extract is also added in bath water for fair skin.

176. *Rosa multiflora* Thunb. (Rosaceae) Pandhra-gulab

Medicine

(*i*) Leaves and flowers are together used to prepare decoction. About 15-20 ml of it is drunk daily to treat leucorrhoea till cure.

177. *Rottboelia cochinchinensis* (Lour.) W.D. Clayton

Miscellaneous Uses

(*i*) It is used as fodder for cattle.

178. ***Saccharum officinarum*** **L. (Poaceae) Sugar-cane**

Medicine

(*i*) Stem juice is consumed for three days to cure hepatitis.

(*ii*) Jaggery is applied around ears to stop hiccup.

(*iii*) Stem juice is boiled till it becomes thick. It is applied on stomach especially around naval to deworm intestinal worm daily till cure.

179. ***Santalum album*** **L. (Santalaceae) Chandan**

Medicine

(*i*) Stem paste is applied on face for eight days against pimples.

180. ***Sapindus emarginatus*** **Vahl (Sapindaceae) Ritha**

Medicine

(*i*) Fruit walls of this species and leaf juice of *Vitex nigundo* L. are homogenized. It is applied on eye injuries.

(*ii*) Fruit paste prepared in water is applied on scorpion-sting.

181. ***Securingea virosa*** **(Roxb. ex Willd.) Baill. (Euphorbiaceae)**

Miscellaneous Uses

(*i*) It is used as fodder for goats

182. ***Semecarpus anacardium*** **L.*f*. (Anacardiaceae) Biba**

Medicine

(*i*) Decoction of fresh stem-bark, about a cup is, mixed with some milk and sugar. It is drunk for three consecutive days to treat rheumatism. Patient is advised to sleep in corner of a house on ground after wrapping himself perfectly. Then the patient on second day should sleep in the middle part of a house. On third day, he is advised to sleep in fore-front of a house.

Miscellaneous Uses

(*i*) Wood is used as fuel and for furniture.

(*ii*) Wood is also employed for roof on huts.

183. *Sesamum orientale* L. (Pedaliaceae) Til

Medicine

(*i*) Seed oil and common salt are mixed. It is then applied on gums daily to treat tooth-ache till cure.

184. *Sesbania bispinosa* (Jacq.) Slend. ex Wight (Fabaceae) Jangali-shevari

Miscellaneous Uses

(*i*) It is used as fodder for goats.

(*ii*) Stems are used for thatching huts.

185. *Setaria verticillata* Beauv. (Poaceae)

Miscellaneous Uses

(*i*) It is used as fodder for cattle.

186. *Sida spinosa* L. (Malvaceae)

Miscellaneous Uses

(*i*) It is used as fodder for goats.

187. *Solanum nigrum* L. (Convolvulaceae) Kamuni

Medicine

(*i*) Decoction of leaves, about a cup, is drunk before tea for two days to cure intestinal worm till cure.

Miscellaneous Uses

(*i*) The stems are used as fodder for pet animals.

(*ii*) Ripe fruits are consumed.

188. *Solanum virginianum* L. (Solanaceae) Ran-wang, Bhui-ringani

Medicine

(*i*) Few roots pieces are wrapped in the leaves of *Piper betle* L. It is consumed for 15 days to cure viral fever and typhoid.

(*ii*) Extract of fruit juice, 2-3 drops, is applied once for three days to cure eye burning.

189. *Spermadictyon suaveolens* Roxb. (Rubiaceae) Padri

Medicine

(*i*) Bark is dried and burnt. The ash is applied directly on a wound daily to check itching till cure.

190. *Spilanthus calva* DC. (Asteraceae) Akkalkadha

Medicine

(*i*) Leaves are chewed thrice a day for three days to cure cough.

191. *Sporobolus coromandeljanus* (Retz.) Kunth (Poaceae)

Miscellaneous Uses

(*i*) It is used as fodder for cattle.

192. *Striga asiatica* (L.) O. Ktze. (Scrophulariaceae) Taluka

Medicine

(*i*) Entire plant is crushed. The extract, about half glass, is drunk for abortion for 5-7 days.

(*ii*) Plant juice is applied on stomach to control intestinal worm.

193. *Stylosanthes fructiosa* (Retz.) Alst. (Fabaceae) Stylo

Miscellaneous Uses

(*i*) It is used as fodder for cattle.

194. *Syzigium cumini* (L.) Skeels (Myrtaceae) Jambhul

Medicine

(*i*) Seeds are dried and powdered. One spoonful twice a day with milk, honey or water is administered against diabetes.

(*ii*) Seeds of this species are crushed along with the seeds of 'kadujeera' [*Vernonia anthelmintica* (L.) Willd.] and five leaves of neem (*Azadiracta indica* A. Juss. and fruit juice of *Momordica charantia* L.). They are homogenized. Pellets are prepared one pellet day is consumed, for 15 days to treat diabetes.

195. *Tamarindus indica* L. (Caesalpiniaceae) Chinch

Medicine

(*i*) Seed coats are removed and the cotyledonous are made into paste. It is applied against scorpion-sting.

Miscellaneous Uses

(*i*) Leaves are used as fodder for goats and sheep.

(*ii*) Wood is used as fuel.

196. *Tectona grandis* L. *f.* (Verbenaceae) Sag

Medicine

(*i*) Seed paste is applied around navel to treat stomach-ache.

197. *Terminalia arjuna* (Roxb.) Wight and Arn. (Combretaceae) Arjun-sadada

Medicine

(*i*) Bark of trunk is heated on a pan. It is powdered and sieved. It is homogenized with coconut oil (*Coccous nucifera* L.) and applied on wounds thrice a day till cure.

(*ii*) Bark is powdered finely. A spoonful of powder is mixed with a spoonful of honey. A pellet prepared is consumed before tea to cure wound till cure.

(*iii*) A spoonful bark powder is consumed with honey or milk once a day against heart complaints for a week.

Miscellaneous Uses

(*i*) Wood is used for furniture purposes and also for huts and houses.

(*ii*) Wood is employed to make wheels of bullock-cart.

(*iii*) The twigs are used for thatching huts.

198. *Terminalia bellirica* (Gaertn.) Roxb. (Combretaceae) Behda

Medicine

(*i*) Seeds are crushed and soaked in water. About 10ml of it is drunk to cure cough. It is advised for two times daily for three consecutive days.

Ethnoveterinary Medicine

(*i*) Crushed seeds are powdered finely. About 10gm of it's mixed in wheat flour. It is fed for four days to pet animal's wormicide for intestinal worms.

Miscellaneous Uses

(*i*) Leaves are used as fodder for goats.

(*ii*) The wood is used to make bullock-carts and also as fuel.

199. *Terminalia catappa* L. (Combretaceae) Badam

Miscellaneous Uses

(*i*) Leaves are used as fodder for cattle.

(*ii*) Wood is used as small timber for various purposes.

200. *Terminalia chebula* Retz. (Combretaceae) Hirda

Medicine

(*i*) Few drops of leaf juice are applied on eyes to cure eye-wounds.

201. *Themeda quadrivalvis* Ktze. (Poaceae) Nil

Miscellaneous Uses

(*i*) It is used as fodder for cattle.

202. *Thespesia lampas* (Cav.) Dalz. (Malvaceae) Ranbhendi

Medicine

(*i*) Roots are crushed. Then decoction is prepared, two spoons of it are administered once for eight days to cure piles.

Miscellaneous Uses

(*i*) The stems are used to make partition walls of huts.

203. *Tinospora cordifolia* (Willd.) Miers. (Menispermaceae) Gulwel

Medicine

(*i*) Leaf extract is diluted in water. Two spoonful of it is administered once a day as antipyretic. It is advised till cure.

(*ii*) Extract of leaves is drunk. One spoonful twice a day is advised to cure cough for 3-4 days.

(*iii*) Decoction of green fruits is prepared. A cup of it is drunk for two days to cure fever and liver complaints.

204. *Trachyspermum ammi* (L.) Sprague (Apiaceae) Owa

Medicine

(*i*) Seeds are consumed daily after meal to improve digestion. It is usually advised during post-delivery period.

(*ii*) Cigar filled with the seeds is smoked (inhaled) twice a day to cure cough.

205. *Tribulus lanuginosus* L. (Zygophyllaceae) Gokharu

Medicine

(*i*) Entire plants are dried and powdered. A spoonful powder once a day is consumed with honey as a tonic for 15 days.

(*ii*) Fruits of this plant and seeds of *Foeniculum vulgare* Mill are powdered. Two spoons in water are drunk to cure stomach-ache.

Ethnoveterinary Medicine

(*i*) Leaf juice is extracted and 10-15 ml of it is dropped into ears to cure worm till cure.

206. *Trichodesma zeylanicum* (Burm.*f.*) R.Br. (Boraginaceae) Nil

Miscellaneous Uses

(*i*) Leaves are rubbed on tongue of pet animals when they do not consume fodder.

207. *Tridax procumbens* L. (Asteraceae) Kambarmodi, Tukdi

Medicine

(*i*) Leaf extract is mixed in ghee and honey. About 2-3 spoons of it is drunk to check passing of blood through urination. It is advised till cure but administered before tea.

(*ii*) Extract of leaf is applied directly to cure wounds.

(*iii*) Leaf juice is gargled to check bleeding of gums.

(*iv*) A cup of decoction is drunk before tea for three to four days to cure wounds and to check fever.

208. *Trigonella foenum-graecum* L. (Fabaceae) Methi

Medicine

(*i*) Seeds powder, about 10 gm, is consumed once a day before tea to cure diabetes. It is advised till cure.

(*ii*) Apart from leafy vegetable its consumption improves appetite. It is therefore also advised.

(*iii*) Decoction of leaves is drunk to cure rheumatism. Half cup of it is administered daily at night.

209. *Tylophora indica* (Burm. *f.*) Merr. (Asclepiadaceae) Sadamas, Satax

Medicine

(*i*) Roots are dried and powdered finely. It is mixed in honey and pellets are prepared. Two pellets are consumed for 21 days to treat stammering.

210. *Vitex negundo* L. (Verbenaceae) Nirgudi

Medicine

(*i*) Juice of fresh leaves mixed in latex of *Calotropis procera* (Ait) R. Br. and some butter. It is applied on eyes to treat eye infection.

(*ii*) Leaf extract (few drops) homogenized with some butter. Two to three drops of it are poured into nose as a remedy against migraine.

(*iii*) Paste of leaves of *Azadirachta indica* A. Juss. + *Pongamia pinnata* (L.) Pierre is prepared. It is applied on joints to cure rheumatism and also sprains.

Miscellaneous Uses

(*i*) The stem is used for making roof, partition walls of huts and houses.

211. *Wattakaka volubilis* (L.*f*.) Stapf. (Asclepiadaceae) Kawali

Medicine

(*i*) Small root pieces are dipped in tea. It is then drunk to get relief from pains of piles.

212. *Withania somnifera* (L.) Dunal (Solanaceae) Aajkand, Ashwagandha

Medicine

(*i*) Roots are dried and powdered finely. It is mixed in milk with of an egg and some oil to prepare paste. It is applied twice a day on body to massage to cure paralysis.

(*ii*) Root pieces are consumed with meal to cure physical weakness. It is advised for a fortnight. Root powder + dry fruit (in 1:1 ratio) are used to prepared sweet balls. One sweet-ball is consumed per day for 30 days to increase sexual excitement.

(*iii*) Roots and dried leaves are powdered. It is mixed in honey and consumed as brain tonic. Two spoons are advised per day.

213. *Woodfordia fructicosa* (L.) Kurtz (Lythraceae) Dhayti

Medicine

(*i*) Flowers are extracted. The juice, 2-4 spoons, of it is administered as a remedy against to improve irregular urination.

214. *Zingiber officinale* Rosc. (Zingiberaceae) Adrak

Medicine

(*i*) Pieces of rhizome are added while preparing tea. It is advised for better digestion and to treat flatulence.

(*ii*) Extract of rhizome is mixed with honey and drunk against cough. Two spoons of it are advised daily till cure.

215. *Ziziphus mauritiana* Lamk. (Rhamnaceae) Bor

Medicine

(*i*) Leaves are crushed in urine of a cow. It is applied directly on hair at night for eight days to prevent hair loss.

Miscellaneous Uses

(*i*) The twigs are used for fencing field and farmyards.

(*ii*) The leaves are fed to goats.

216. *Ziziphus nummularia* (Burm.*f.*) Wight and Arn.: (Rhamnaceae)

Miscellaneous Uses

(*i*) It is used as fodder for goats.

SECTION–IV
(*Lonar, Sindkhed Raja & Dewulgaon Raja Tehsils*)

1. ***Abelmoschus manihot*** **(L.) Medik. ssp.** ***tetraphyllus*** **(Roxb. ex Horn.) Borssum (Malvaceae) Ran-bhendi**

Medicine

(*i*) Young raw fruits (3-4) are consumed alongwith honey daily once for three to four days to treat cough.

2. ***Abrus precatorius*** **L. (Fabaceae) Lal-gunj (Red seeded variety), Safed-gunj (White seeded variety)**

Medicine

(*i*) Leaves are first washed and then consumed to clear throat and voice.

(*ii*) Stem powder (3-4 mg) is mixed in honey and consumed daily for 3-4 days to cure cough.

(*iii*) Leaves and roots are crushed together. This paste, about 5 gm, is consumed daily to treat chronic cough till cure.

(*iv*) Consumption of leaves is also advised for mouth-ulcer. Few leaves are chewed for 3-4 times a day till cure.

(*v*) Green pods (White seeded variety) are extracted in water. This extract, about 20 ml, is drunk to treat leucorrhoea till cure.

3. *Abutilon indicum* (L.) Sweet (Malvaceae) Dabel, Dabala

Medicine

(*i*) Extract of few fresh leaves is mixed in cow-ghee in equal quantity. About two spoons of it are administered daily for three days to regularise menstruation.

Miscellaneous Uses

(*i*) Leaves are used as fodder for cattle.

4. *Acacia chundra* Willd.(Mimosaceae) Khair

Medicine

(*i*) Stem and leaves are burnt. The ash is mixed in coconut oil (*Cocos nucifera* L.). It is applied daily on scabies, burns and wounds till cure.

5. *Acacia nilotica* (L.) Willd. ex Del. subsp. *indica* (Bth.) Brenan (Mimosaceae)Babhul:

Medicine

(*i*) Flowers are crushed in cow-ghee or cow-milk and spoonful of it is consumed at morning and evening to cure jaundice till cure.

(*ii*) Leaf extract and common salt are homogenised. It is applied daily on injury of legs till cure.

(*iii*) Stem bark powder is boiled in water. This decoction is gargled for two to three times a day to cure mouth ulcer.

(*iv*) Young leaves are chewed to treat throat infection for few days.

(*v*) Fresh flower paste is applied on body of a child suffering from chicken-pox.

6. *Acacia nilotica* (L.) Del. subsp. *cupressiformis* (Stewart)Ali and Faruqui (Mimosaceae)Vedi-babhul

Medicine

(*i*) Young leaves are crushed. About 15-20 ml of this extract is drunk at morning to reduce body heat. This is also advised for treating piles.

7. *Acalypha indica* L. (Euphorbiaceae) Tankhalan

Medicine

(*i*) Leaf extract (in 1:1 ratio) is mixed in curd. A spoon of it is drunk to treat jaundice. It is administered for seven days.

(*ii*) Leaf paste is homogenised with jaggery. It is applied for gangrene till cure.

(*iii*) Leaf paste is applied on body of children and two to three drops are advised orally daily to cure colic problem till cure.

(*iv*) Leaf extract, about half cup, is taken orally for three days at morning and evening to control dysentery.

Ethnoveterinary Medicine

(*i*) Leaf extract is applied on wounds daily in case of bullocks till cure.

8. *Achyranthes aspera* L. (Amaranthaceae) Aghada, Zinzoda

Medicine

(*i*) Two leaves are crushed alongwith common salt. This paste is daily applied on infected teeth till cure.

(*ii*) Leaf juice, one spoonful twice daily, is drunk to cure piles till cure. Leaf paste is also applied on injuries caused due to piles.

(*iii*) Half spoon of seed powder is dissolved in water. It is drunk daily to cure cough till cure.

(*iv*) Root extract is mixed in human urine in equal proportion. A cup of it is administered to a person suffering from snake-bite.

Ethnoveterinary Medicine

(*i*) Root extract, about two to three cups, is given orally to a calf to control dysentery.

9. *Adansonia digitata* L. (Bombacaceae) Gorakh-chinch

Medicine

(*i*) Leaves or bark are boiled and half cup of it is drunk once daily to cure dysentery, sunstroke or colic problem till cure.

10. *Adiantum philippense* L.(Adiantaceae) Rajhans

Medicine

(*i*) Leaves are crushed in water and one or two drops are applied on eyes to prevent conjunctivitis till cure.

11. *Aegle marmelos* (L.) Corr. (Rutaceae) Bel

Medicine

(*i*) Fruit paste is applied on knees to reduce rheumatic pain till cure.

(*ii*) Leaves are added while preparing tea. This tea is drunk at early morning to control dysentery till cure.

12. *Aerva lanata* (L.) Juss. ex Schult. (Amaranthaceae) Hatti-sondi

Medicine

(*i*) Leaf juice is mixed in curd in 1:1 ratio. It is consumed, about twenty ml daily, at morning as a remedy against kidney-stone till cure.

13. *Ailanthus excelsa* Roxb. (Simaroubaceae) Waghat, Mahapurush

Medicine

(*i*) Dried fruit powder, about spoonful, is consumed for seven days to treat diabetes.

(*ii*) Leaf juice is mixed with cow-ghee. It is applied on different skin diseases.

Ethnoveterinary Medicine

(*i*) Leaf juice is mixed in sulphur powder. Fruit powder of Awala (*Emblica officinalis* Gaertn.) is also mixed in it. This mixture is prepared in a copper plate. It is then applied around horns of a pet animal to kill worms. It is also applied on neck for neck-ache.

14. *Allium cepa* L. (Liliaceae) Kanda

Medicine

(*i*) Onion is cut and it is smelt in case of problem of epilepsy.

(*ii*) Extract of bulb is mixed in cow-ghee in 1:1 ratio. A spoonful of it is drunk daily at morning to reduce acidity till cure. It is also beneficial in case of swollen stomach.

(*iii*) Bulb paste is homogenized in honey. It is applied on body to treat rheumatism till cure.

(*iv*) Extract of onion bulb and ginger (*Zingiber officinale* Rosc.) are mixed in equal quantity. One or two spoons of it are advised once to avoid vomitting.

(*v*) About 3-4 ml of bulb extract is administrered only once to a child suffering from intestinal pains.

15. *Allium sativum* L. (Liliaceae) Lasun

Medicine

(*i*) Three bulblets are crushed and this paste is boiled slightly in coconut oil (*Cocos nucifera* L.). Few drops of it are poured into ear to treat ear-ache or to prevent pus-formation.

(*ii*) Three bulblets are boiled in a cup of water. It is drunk daily once to control cough till cure.

(*iii*) Five bulblets are boiled in a cup of water. It is drunk. Some fruits of chilly (*Capsicum annuum* L.) and common salt are crushed and paste is prepared. It is fried in oil. This paste, about three spoons, is consumed daily for eight days to treat indigestion.

(*iv*) Few bulbs are made into paste. It is applied on head to control head-ache.

16. *Aloe vera* L.(Liliaceae) Korphad

Medicine

(*i*) One or two pieces of leaf pulp alongwith betel leaf (*Piper betle* L.) are consumed twice or thrice daily for five days to cure piles.

(*ii*) Leaf pulp about 5 gm only once is consumed as purgative.

(*iii*) Leaf pulp (10 gm) and some candy (*Saccharum officinarum* L.) are homogenised. It is consumed daily to treat sunstroke and scant urination till cure.

(*iv*) Leaf pulp is boiled in cow-ghee. About half cup of it is consumed daily to cure piles till cure.

17. *Amaranthus spinosus* L.(Amaranthaceae) Katya-math

Medicine

(*i*) Leaf juice, about spoonful, is advised for children at morning and evening to cure pneumonia till cure.

(*ii*) Root paste is applied to cure red eyes in case of children.

(*iii*) Leaf paste is applied once daily on wounds till healing.

18. *Amorphophallus commutatus* (Schott.) Engl. (Araceae) Kolha-kand

Medicine

(*i*) Rhizome paste is applied on body part after scorpion-sting.

(*ii*) Rhizome is extracted. It is mixed in coconut oil (*Cocos nucifera* L.). It is heated and applied for 3-4 days at morning to treat hair-blisters.

19. *Ampelocissuus latifolia* (Roxb.) Planch. (Vitaceae) Mhaiswel

Medicine

(*i*) Leaf juice is applied daily on blisters till cure.

20. *Annona squamosa* L.(Annonaceae) Sitafal

Medicine

(*i*) Roots are cut into pieces. These are strung around waist of a woman before delivery for smooth delivery.

21. *Anogeissus latifolia* (Roxb. ex DC.) Wall. ex Guill. (Combretaceae) Dhavda

Medicine

(*i*) Seeds of Dhavda, Behda, [*Terminalia bellirica* (Gaertn.) Roxb.] and leaves of Awala (*Emblica officinalis* Gaertn.) are crushed together. This extract, about two spoons twice daily, is advised for 2-3 days to control fever.

22. *Argemone mexicana* L. (Papaveraceae) Bilait

Medicine

(*i*) Root paste is applied on tooth ache.

(*ii*) Root paste is also applied daily on piles till cure.

(*iii*) Root paste is applied on scorpion sting or root juice (about 20 ml) is also taken orally to treat scorpion sting.

(*iv*) Root paste alongwith betel leaf (*Piper betle* L.) is consumed for three to four days thrice a day to treat malarial fever.

(*v*) Root juice (about 20 ml) is also taken orally to treat scorpion sting.

23. *Aristolochia bracteolata* Retz. (Aristolochiaceae) Gidhad

Medicine

(*i*) Root pieces are deeped in a cup of water for about fifteen minutes. This infusion is administered to a person suffering from snake-bite.

(*ii*) Leaf extract is applied on wounds to deworm wounds. It is applied once daily for three to four days.

(*iii*) Leaf extract, two to three drops, are poured once daily into ears to prevent pus-formation till cure.

24. *Azadirachta indica* A. Juss. (Meliaceae) Nimb

Medicine

(*i*) Leaf extract is applied daily once on body to check fever till cure.

(*ii*) A cup of leaf extract is drunk for three days at morning and at night to treat diarrhoea.

(*iii*) A twig cut from the tree is kept erect in water for sometime. It is then extracted. A cup of it is advised for 8-10 days at morning and night to purify blood.

(*iv*) Stem bark powder is sprinkled daily on body to cure small-pox till cure.

(*v*) Stem bark paste, about four to five gm, is consumed twice a day for three consecutive days to cure fever.

(*vi*) Stem bark is extracted in water. Spoonful of it is administered daily once to control cough till cure.

25. *Bacopa monnieri* (L.)Wettst. (Scrophulariaceae) Bam, Nili

Medicine

(*i*) Leaf extract is mixed in milk 1:2 ratio. A cup of it is drunk before meal to treat leucorrhoea till cure.

(*ii*) Seeds are crushed. The extract is applied on head as hair tonic for a fortnight.

26. *Balanites aegyptiaca* (L.) Del. (Balanitaceae) Hingan-bet

Ethnoveterinary Medicine

(*i*) Seed paste is applied in the eyes of a bullock to clear eyes when they turn white.

27. *Barleria prionitis* L. (Acanthaceae) Katya-kul, Kate-koranti

Medicine

(*i*) Leaves of this plant, rhizome of ginger (*Zingiber officinale* Rosc.) and catechu (*Acacia catechu* Willd.), lime and three cloves (*Syzygium aromaticum* Merr. and Perry) are wrapped in a betel leaf (*Piper betle* L.). It is consumed daily to treat typhoid till cure.

(*ii*) Leaf paste is applied on lumped part till cure.

(*iii*) Leaf paste is mixed in coconut oil (*Cocos nucifera* L.) and applied on wounds and scabies till cure.

(*iv*) Few pieces of roots wrapped in a betel leaf (*Piper betle* L.). It is comsumed daily twice a day to cure piles.

28. *Bauhinia racemosa* Lamk. (Caesalpiniaceae) Apta

Medicine

(*i*) Leaf powder is mixed into honey. One spoonful of it is consumed at morning to cure asthma and tuberculosis.

29. *Biophytum sensitivum* (L.) DC. (Oxalidaceae) Lajalu

Ethnoveterinary Medicine

(*i*) Leaf paste is applied daily once on wounds of cows and bullocks till cure.

30. *Blepharis repens* (Vahl) Roth (Acanthaceae) Hadsan

Medicine

(*i*) Entire plants are washed and dried. These are powdered. It is mixed in sweetmeat and consumed at morning to cure bone-fracture. It is followed till cure.

31. *Boerhavia diffusa* L. (Nyctaginaceae) Safed-vasu, Vasu

Medicine

(*i*) Jaggery is mixed in leaf juice. It is applied on body to treat small-pox for few days.

(*ii*) Leaf extract, about 10 ml, is beneficial in controlling jaundice. It is advised daily once till cure.

(*iii*) Similarly, root extract is mixed in fresh cow's milk in 1:1 ratio. It is drunk as a cure against jaundice. About twenty ml of it is advised daily once to cure.

(*iv*) Pieces of stem bark are strung around neck of a patient suffering from jaundice till cure.

(*v*) Roots paste is mixed in a cup of milk. It is drunk twice daily to cure jaundice till cure.

32. *Buchanania lanzan* Spreng. (Anacardiaceae) Char

Medicine

(*i*) Gum is slightly diluted in water. It is applied daily on joints to treat arthritis till cure.

33. *Butea monosperma* (Lamk.)Taub. (Fabaceae) Palas

Medicine

(*i*) Petiole of fresh leaf is chewed by adults. It is so extracted and swallowed once a day to treat acidity till cure.

(*ii*) Leaves are crushed in castor oil (*Ricinus communis* L.). This is drunk, about 10 ml, daily to check rheumatism.

(*iii*) Flowers are boiled and wrapped in cloth. This is used to warm abdomin to control stomach-ache.

(*iv*) Leaves are warmed and wrapped around joints to treat rheumatism till cure.

Ethnoveterinary Medicine

(*i*) Flowers are boiled and wrapped in cloth. Stomach of pet animal is warmed using it to treat stomach-ache.

34. *Caesalpinia bonduc* (L.) Roxb. (Caesalpiniaceae) Sagargoti

Medicine

(*i*) Seeds are roasted, then powdered and made into paste. Pellets are prepared. One pellet is advised daily to cure intestinal ulcer. It is followed for 4-5 days.

(*ii*) Seeds are powdered. It is mixed in equal quantity of sugar (*Saccharum officinarum* L.). It is consumed daily at morning and evening to treat leucorrhoea. Tea should be avoided during the treatment.

35. *Caesalpinia decapetala* (Roth) Alst. (Caesalpiniaceae) Katyawel

Medicine

(*i*) Equal number of seeds of this plant and those of sagargoti (*Caesalpinia bonduc* L.) Roxb. and Hinganbet [*Balanites aegyptiaca* (L.) Del.] are powdered. A spoonful of this powder dissolved in water is consumed once daily to treat stomach-ache till cure.

36. *Caesalpinia pulcherrima* (L.) Swartz (Caesalpiniaceae) Shankasur

Medicine

(*i*) Leaves are cooked with rice. This is consumed daily to cure rheumatism for a week or so.

37. *Cajanus cajan* (L.) Millsp. (Fabaceae) Tur

Medicine

(*i*) Leaf paste is applied on cuts and wounds till cure.

Ethnoveterinary Medicine

(*i*) About half kg of pulse is cooked in half litre of water. It is fed to bullocks and cows to treat watery diarrhoea. This is advised daily till cure.

38. *Calotropis gigantea* (L.) R. Br. (Asclepiadaceae) Pandhri-Ruchkin, Rui

Medicine

(*i*) Leaf latex is diluted slightly in water. Few drops are dropped into nose to cure head-ache.

(*ii*) Central part of flowers is consumed once a day to treat cough till cure.

(*iii*) One flower alongwith a betel leaf (*Piper betle* L.) is consumed twice a day to treat cough till cure.

(*iv*) A pinch of vermilion powder and few flowers are made into paste. Paste is prepared in castor oil (*Ricinis communis* L.). It is applied around ears to treat ear-ache.

(*v*) Leaf latex and groundnut oil (*Arachis hypogaea* L.) and 'shendur' (Red oxide) are homogenised. This mixture applied on eczema.

(*vi*) Flowers are wrapped in a betel leaf (*Piper betle* L.) and consumed at morning and evening to cure whooping cough.

(*vii*) Stem pieces are smoked to cure head-ache.

(*viii*) Few flowers are wrapped in a betel leaf (*Piper betle* L.) and consumed once for three days to get relief from asthma.

(*ix*) Flower juice (about 5 ml) is taken orally at morning and evening for 3-4 days to get relief from asthma.

(*x*) Leaf latex is mixed in soil. It is then applied on swelling of legs till cure.

Ethnoveterinary Medicine

(*i*) Leaf latex is mixed in two spoonful oil of sunflower (*Helianthus annuus* L.) It is applied in bullock's eyes for better vision.

39. *Cannabis sativa* L. (Cannabinaceae) Ganja

Medicine

(*i*) Seeds are crushed and mixed in cow urine. It is heated and then used to cure to piles. This gives relief.

40. *Canthium coromandelicum* (Burm *f.*) Alston (Rubiaceae) Fetari

Medicine

(*i*) Leaf extract is mixed in jaggery. Pellets are prepared. Two pellets for few days are advised to treat dog-bite.

41. *Capparis grandis* L.*f.* (Capparidaceae) Thikari

Medicine

(*i*) Decoction of stem bark and leaves is applied daily once on joints to treat rheumatism till cure.

42. *Capparis zeylanica* L. (Capparidaceae) Waghat-wel

Medicine

(*i*) Leaf extract, about 15-20 ml, is administered for seven days to a woman suffering from leucorrhoea for a week.

43. *Capsicum annuum* L. (Solanaceae) Mirchi

Medicine

(*i*) Leaf juice, 2-3 drops, is poured into nose to treat head-ache once a day till cure.

44. *Cardiospermum halicacabum* L. (Sapindaceae) Kapalphuti

Ethnoveterinary Medicine

(*i*) Leaf juice is extracted. About a litre of it is administered to bullocks orally to treat dysentery for two or three days.

(*ii*) Leaf juice, 1-2 litres, is also advised for bullocks orally to cure flatulence.

45. *Careya arborea* Roxb. (Lecythidaceae) Khumbai

Medicine

(*i*) Fruit are extracted. About two spoons of it is drunk. It is beneficial for stomach-ache. Only one dose is daily advised.

Ethnoveterinary Medicine

(*i*) Fruits are crushed and its paste is applied on wounds of cows, buffaloes and bullocks..

(*ii*) Fruits are crushed and extracted in water. This extract, about 3-4 glasses, is administered as a remedy against flatulence in case of cows and bullocks.

(*iii*) Fruit powder is mixed in water. It is administered to a bullock suffering from blood dysentery. Two to three fruits at a time are used. This is followed for three to four days.

46. *Carica papaya* L. (Caricaceae) Papai

Medicine

(*i*) Fruit latex is applied daily on injury and also on piles till cure.

(*ii*) Pulp of unwripe fruits is homogenised with some common salt. About 10-15 gm of it is consumed to treat kidney-stone for a week.

(*iii*) Pulp of raw fruit is consumed to treat stomach-ache.

47. *Cassia auriculata* L. (Caesalpiniaceae) Tarwad, Awali

Medicine

(*i*) Flower paste is prepared in water. It is applied on redness of eyes.

(*ii*) Fruits are powdered. A spoonful of it is consumed twice daily to treat asthma till cure.

48. *Cassia fistula* L. (Caesalpiniaceae) Bahava

Medicine

(*i*) Pod is dried and powdered. About 3-4 gm is dissolved in water and drunk twice a day to cure inflammation of body or any organ. Common salt and pungent substances should be avoided.

49. *Cassia obtusifolia* L. (Caesalpiniaceae) Tarota, Tarvad

Medicine

(*i*) Leaf juice, about 25 ml, is drunk at morning before tea for jaundice and rheumatism for a week.

(*ii*) Leaves wrapped in a cloth are tied at night on legs to get relief from rheumatism.

(*iii*) Leaf juice, about a glass, is advised to drink against dog-bite and wolf-bite.

(*iv*) Half cup of leaf juice is advised at morning during pregnancy for better health.

(*v*) Leaf juice and cow-ghee is mixed and drunk for three days during pre- and post-delivery period to increase disease resistance.

(*vi*) Seed powder (about 100 gm) is consumed daily for a week for abortion.

(*vii*) Seed powder is also advised (about 5 gm) daily for seven days to reduce rheumatic pains.

(*viii*) Leaves are boiled and then crushed. This paste is applied at night daily to treat arthritis till cure.

50. *Cassia tora* L. (Caesalpiniaceae) Raj-tarota

Medicine

(*i*) Leaf paste is applied daily on scabies and ringworm till cure.

(*ii*) Leaves are wrapped in cloth. This is covered by ash. It is then heated. The leaves are extracted when hot. Half cup of this extract is daily drunk for a fortnight as a remedy against rheumatism.

51. *Cassine alberns* (Retz.) Kosterm. (Celastraceae) Bhuti-kalas

Medicine

(*i*) Leaves are powdered. A pinch of powder is dissolved in spoonful water. It is advised daily for three-four times only at the interval of 2-3 hours for a child suffering from dysentery and vomiting.

52. *Catharanthus roseus* (L.) G. Don (Apocynaceae) Sada-fuli, Sadapa

Medicine

(*i*) Leaf juice is applied on a child. Few drops are also advised orally only once to treat vomiting and dysentery.

(*ii*) Fresh 3-4 flowers are consumed at morning before food to treat diabetes for a week.

53. *Celastrus paniculata* Willd. (Celastraceae) Malkangani

Medicine

(*i*) Seeds are crushed and oil is obtained. It is applied on joints to control rheumatic pains. It is also applied on paralysed organs till cure.

54. *Celosia argentea* L. (Amaranthaceae) Kurdu

Medicine

(*i*) Root paste, about 10 gm, mixed in water is consumed at morning and evening to cure kidney-stone for a week.

55. *Chenopodium murale* L. (Chenopodiaceae) Son-gawat

Miscellaneous Uses

(*i*) It is used as fodder.

56. *Cissus quadrangula* L. (Vitaceae) Kandwel

Medicine

(*i*) Petiole paste is applied on nails to cure injuries.

57. *Citrullus colocynthis* (L.) Schrad. (Cucurbitaceae) Kadunar, Kaduindravan

Medicine

(*i*) Leaf extract, 5-10 ml, is advised to drink twice a day to treat fever till cure.

(*ii*) Leaf and roots (few pieces) of this plant are wrapped in a betel leaf (*Piper betle* L.). This is consumed daily once as a remedy against stomach-ache.

(*iii*) Root paste, about 2-3 gm, is consumed daily once to treat acidity daily at night till cure.

(*iv*) Decoction of roots is applied daily at night on head for better hair growth.

58. *Citrus aurantifolia* (Christm.) Sw. (Rutaceae) Gal-limbu

Medicine

(*i*) Oil is obtained from fruit rinds. About twenty to twenty five ml oil is drunk if any metal body like coins is swallowed. This helps expulsion from intestine.

(*ii*) Fruit juice is extracted. A cup of it is taken orally for three to four days to treat kidney-stone.

(*iii*) Fruits are boiled in water. This water is taken orally in morning to cure diabetes till cure.

(*iv*) Fruit juice is mixed in equal quantity of milk and half cup of it is drunk daily to cure piles till cure.

59. *Clematis gouriana* Roxb. (Ranunculaceae) Ranmogra

Medicine

(*i*) Leaf juice is mixed in equal ratio in coconut oil (*Cocos nucifera* L.). It is applied on gums to check tooth-ache.

60. *Clerodendrum inerme* (L.) Gaertn. (Verbenaceae) Koynel, Shendge Ghashyache- Pan

Medicine

(*i*) Young leaves and common salt are consumed if any object is obstructed in alimentary canal. This is advised till the obstruction is removed.

61. *Cleodendrum multiflorum* (Burm.*f*.) O. Ktz. (Verbenaceae) Kalitankhalan

Medicine

(*i*) Root and leaf extract is applied daily on joints to treat arthritis till cure.

(*ii*) Leaf paste is mixed with candy (*Saccharum officinarum* L.). It is applied on gangrene and blisters as well.

(*iii*) Leaf juice is applied on body of a child and two to three drops are advised to drink to cure colic.

Miscellaneous Uses

(*i*) Leaves are used as fodder for goats.

62. *Cleome viscosa* L. (Capparidaceae) Pivali-til

Medicine

(*i*) Two to three drops of leaf juice are poured daily once into nose to treat head-ache till cure.

63. *Clitoria ternatea* L. (Fabaceae) Gokarna-wel

Medicine

(*i*) Leaf extract is, 4-5 drops, is added in mother's milk. It is advised only once for a baby for stomach ache.

64. *Coccinia grandis* (L.) Voigt (Cucurbitaceae) Tondali

Medicine

(*i*) Root paste is applied daily on abscess.

Ethnoveterinary Medicine

(*i*) Raw fruit, about 500 gm, are crushed in half litre of water. This is fed at morning to the cattle for seven days to treat dizziness.

65. *Cocculus hirsutus* (L.) Diels. (Menispermaceae) Wasant-wel, Wasan

Medicine

(*i*) Leaf powder and candy (*Saccharum officinarum* L.) in equal quantity are mixed. It is homogenised in whey and consumed twice daily to reduce pains after menopause.

(*ii*) Fresh leaf juice, about two spoons mixed in equal quantity of cow-ghee, is daily drunk to regularise menstruation.

66. *Cocos nucifera L.* (Arecaceae) Naral

Medicine

(*i*) Kernel is chewed daily for 4-5 days to cure mouth ulcer.

67. *Corchorus depressus* L. (Tiliaceae) Safed-harar

Medicine

(*i*) Root pieces are strung around head to treat malaria till cure.

68. *Corchorus olitorius* L. (Tiliaceae) Choch

Medicine

(*i*) Leaf is consumed as vegetable for some days to treat rheumatic pains of knees and waist.

69. *Cordia dichotoma* Forst.*f.* (Boraginaceae) Bhokar

Medicine

(*i*) Decoction of bark is gargled in case of throat infection daily once till cure.

70. *Cordia gharaf* (Forsk.) Ehrenb. (Boraginaceae) Gondan

Medicine

(*i*) Leaf is extracted in water. A spoonful of it is advised daily for ten days to treat kidney-stone.

(*ii*) A leaf and pinch of common salt is consumed together for 2-3 days to treat cough.

(*iii*) Roots are used as tooth-brush to reduce tooth-ache.

71. *Coriandrum sativum* L. (Apiaceae) Dhane

Medicine

(*i*) Fruits of 'Dhane', some candy (*Saccharum officinarum* L.), few fruits of cardamomum [*Elettaria cordamomum* (L.) Maton] are powdered. A spoonful of it is consumed in water daily once to treat mouth ulcer till cure.

(*ii*) Seeds, some leaves and candy is made into paste. It is, about 3-4 gm, dissolved in cold water and drunk twice daily to cure leucorrhoea till cure.

72. *Crinum latifolium* L. (Amaryllidaceae) Ran-kanda

Medicine

(*i*) Few drops of leaf juice are poured into ears to cure ear-ache for 3-4 days.

73. *Cryptolepis buchanani* R. and S. (Asclepiadaceae) Kavala

Medicine

(*i*) Fruit walls are crushed. It is cooked and mixed in pulse of

Udid [*Vigna mungo* (L.) Hepper]. It is consumed to control stomach ache and flatulence.

Ethnoveterinary Medicine

(*i*) This climber is tied around neck of a bullock suffering from intestinal worms and also for controlling FMD (Foot Mouth Disease).

74. *Cucumis callosus* (Rottl.) Cogn. (Cucurbitaceae) Dangar-wel

Medicine

(*i*) Fruit walls are extracted in water. This extract, about a cup, and about 5 ml decoction of seeds of Pigeon pea [*Cajanus cajan* (L.) Millsp.] is consumed together to treat snake-bite.

75. *Cucumis melo* L. var. *agrestis* Naud. (Cucurbitaceae) Shendadi

Medicine

(*i*) Raw fruits are sliced. Common salt is applied on them. These are consumed daily thrice for few days to treat yellow and scant urination.

76. *Cucumis melo* var. *utilissimus* (Roxb.) Duthie and Fuller (Cucurbitaceae) Kakdi

Medicine

(*i*) Slices of fresh and raw fruits are heated and kept on black facial spots to remove them.

(*ii*) Slices of fruits are also kept on eyes in ease of eye burning.

77. *Cucurbita maxima* Duch.ex Lam. (Cucurbitaceae) Gangafal

Medicine

(*i*) About two spoonful seeds are selected. They are decoated. They are made into paste. This is consumed at night. Two spoons of castor oil (*Ricinus communis* L.) are administered next morning. This helps to kill intestinal worms. This is followed for 4-5 days.

78. *Cucurbita pepo* L. (Cucurbitaceae) Bhopla

Medicine

(*i*) Fruit is smelt after cutting. It is relief for head-ache.

(*ii*) Fruit extract is applied around ears to treat ear-ache.

Ethnoveterinary Medicine

(*i*) Chuna (lime) and paste of fruit are applied on bullock's horns to get relief from intestinal swelling of bullocks.

79. *Cullen corylifolia* (L.) Medik. (Fabaceae) Bawanchi

Medicine

(*i*) Seed oil is applied daily to cure leucoderma.

(*ii*) Stem bark is extracted in water. The extract, about three spoons, is drunk twice daily as a remedy against skin allergy.

80. *Cuminum cyminum* L. (Apiaceae) Jire

Medicine

(*i*) Few leaves or seeds are crushed in honey. It is consumed daily at morning for seven days to treat jaundice.

81. *Curcuma aromatica* Salisb. (Zingiberaceae) Ambi-halad

Medicine

(*i*) Rhizome paste is applied daily on swelling caused due to bone fracture till cure.

(*ii*) Paste of rhizome is mixed in cow-dung. It is applied at evening to cure 'Nagin' disease of skin.

82. *Cuscuta chinensis* Lamk. (Cuscutaceae) Amarwel

Medicine

(*i*) Infusion of stems is used to bath a child for three days to reduce fever.

83. *Cyamopsis tetragonoloba* (L.) Taub. (Fabaceae) Gawar

Medicine

(*i*) Extract is obtained from tender leaves. Few drops are

applied at night on eyes for seven days to cure night blindness.

84. *Cymbopogon citratus* (DC.) Stapf (Poaceae) Gawati-chaha

Medicine

(*i*) Leaves are mixed and boiled in decoction of tea. It is drunk to treat cold.

85. *Cymbopogon martinii* Wats. (Poaceae) Roshya-gawat

Medicine

(*i*) Leaves of this grass and leaves of 'Nilgiri' (*Eucalyptus maculata* Hook.), and a lemon (*Citrus aurantifolia* (Christm.) Sw.) are crushed in coconut oil (*Cocos nucifera* L.). This paste is then applied on knee-joints to treat rheumatism.

(*ii*) Oil is obtained from the fresh leaves. It is applied on legs and hands to cure arthritis till cure.

86. *Dalbergia sisso* Roxb. ex DC. (Fabaceae) Sisam

Medicine

(*i*) Fresh bark is extracted in water. About 4-5 ml of it thrice a day is drunk to control dysentry.

87. *Datura innoxia* Mill. (Solanaceae) Dhotra

Medicine

(*i*) Leaf paste is mixed thoroughly with some common salt. It is applied daily on nails in case of their infection till cure.

(*ii*) Decoction of leaves is prepared. A spoonful of it is drunk twice a day to treat snake-bite.

(*iii*) Leaf paste is homogenised with common salt. It is applied on fingers to treat nail-ache.

88. *Dichanthium annulatum* (Forsk.) Stapf (Poaceae) Butti-gawat

Miscellaneous Uses

(*i*) It is used as fodder for domestic animals.

89. *Dichanthium foveolatum* (Del.) Roberty (Poaceae) Ghas-gawat

Miscellaneous Uses

(*i*) It is used as fodder.

90. *Dinebra retroflexa* (Vahl) Panz. (Poaceae) Ghani-gawat

Miscellaneous Uses

(*i*) It is used as fodder for pet animals.

91. *Dioscorea bulbifera* L. (Dioscoriaceae) Jatashankar

Medicine

(*i*) Dried roots are powdered. It is mixed in water. A spoon of it is drunk twice a day for 40 days to treat rheumatism

(*ii*) Tuber is dried, powdered and consumed to treat stomach-ache. About a spoon of it is consumed once daily for 3-4 days.

(*iii*) Extract of fresh tuber is applied on tumours daily till cure.

(*iv*) Tuber is dried and powdered. The pellets are prepared. Each pellet is consumed alongwith a betel leaf (*Piper betle* L.) at morning, noon and evening to treat colic.

(*v*) Tuber powder is mixed in cow-milk. It is consumed at morning and evening for rheumatism till cure. A glass of it is advised for a day.

(*vi*) Fresh tuber is extracted and mixed in cow-urine. It is applied on wounds and piles.

92. *Diplocyclos palmatus* (L.) Jeffery (Cucurbitaceae) Shivlingi

Medicine

(*i*) Seven seeds are embedded separately in wheat flour (*Triticum aestivum* L.). Thus seven pellets are prepared. One pellet at night is consumed for seven days prior to menstruation period. This helps conceive successfully.

93. *Dolichandrone falcata* (Wall.ex DC.) Seem (Bignoniaceae) Shingi, Melshingi

Medicine

(*i*) Leaf powder in milk, about 10 gm, is consumed at morning to treat piles till cure.

(*ii*) Leaf powder, about one spoon, is consumed for forty days at morning to treat diabetes.

(*iii*) Leaves are heated and tied on legs to cure rheumatism till cure.

94. *Drypetes roxburghii* (Wall.) Hurusawa (Euphorbiaceae) Putran-jiva

Medicine

(*i*) Leaf powder of this plant and those of 'Nimb' (*Azadirachta indica* A. Juss.) and few fruits of *Elettaria cardamomum* Maton are made into paste. This paste is taken orally with cow-ghee to get relief from asthma.

95. *Dunbaria gladulosa* (Dalz. and Gibs.) Prain (Fabaceae) Kach-kuiri

Medicine

(*i*) Fruit hairs are removed. These are crushed and homogenised with jaggery. Pellets are prepared. Two pellets a day are advised to deworm intestinal worms till cure.

96. *Echinops echinatus* Roxb. (Asteraceae) Kate-phal, Utkata

Medicine

(*i*) Roots are extracted. It is slightly diluted in water. It is consumed once to reduce abdominal pains.

(*ii*) Few pieces of roots, a leaf of banyan (*Ficus begalensis* L.) and few date fruits [*Phoenix dactylifera* (L.) Roxb.] and stem pieces of *Tinosopora cordifolia* (Willd.) Miers. are used to prepare decoction. About 5 ml. of it is administered twice daily to treat cough for three days.

(*iii*) Root extract is mixed in milk. About 5 ml is drunk daily once for 5-6 days to check typhoid fever.

97. *Elettaria cardamomum* Maton (Zingiberaceae) Vilaychi, Veldoda

Medicine

(*i*) Root paste is applied onto place of scorpion sting. This paste, about 10 gm, is also advised for consumption.

(*ii*) Root pieces are wrapped in a betel leaf (*Piper betle* L.) and advised to consume for treating malarial fever till cure.

98. *Eleusine coracana* Gaertn. (Poaceae) Nagli

Medicine

(*i*) Flour obtained from seeds is made into paste. This paste is applied on body twice a day for three days to reduce fever.

99. *Emblica officinalis* Gaertn. (Euphorbiaceae) Avala

Medicine

(*i*) Fruits are powdered alongwith sugar (*Saccharum officinarum* L.) and rhizome powder of ginger (*Zingiber officinale* Rosc.) is also mixed in it. About 5 gm of it is consumed before food to treat acidity.

100. *Enicostema axillare* (Lam.) Raynal (Gentianaceae) Nay

Medicine

(*i*) Leaves are boiled. This decoction, about a cup, is drunk at evening daily to control fever till cure.

(*ii*) Leaf juice and some powder of brick are homogenised. It is sieved through cloth. A spoonful of it is twice a day is drunk to cure cough.

(*iii*) Leaf paste is mixed in cow-urine or cow-ghee. It is administered to apply on blisters daily till cure.

(*iv*) Leaf juice is extracted in water. A spoonful of it is advised for children to treat malarial fever.

(*v*) Leaf juice, about 10 ml, to children to control dysentery. It is advised 2-3 times daily till cure.

101. *Erythrina variegata* L. (Fabaceae) Chopda-panghar

Medicine

(*i*) Leaf paste (10 gm) and some candy (*Saccharum officinarum* L.) are dissolved in a cup of whey. It is consumed twice daily for seven days to treat leucorrhoea.

102. *Euphorbia tirucalli* L. (Euphorbiaceae) Shera-thor

Medicine

(*i*) Young green branches are wrapped in cloth. It is heated and then used to warm joints to check arthritis till cure.

(*ii*) Young green branches are extracted in seasame oil (*Sesamum orientale* L.). It is mixed in jaggery. Pellets are prepared. One pellet is advised for few days to cure jaundice.

103. *Ficus benghalensis* L. (Moraceae) Vad

Medicine

(*i*) Latex obtained after cutting leaves is mixed in 'Shendur' (Red lead) mixed in it. It is applied daily on wounds till cure.

(*ii*) Mature leaves are boiled in water. Half cup of this decoction is advised for three days at morning to treat kidney-stone.

104. *Ficus nervosa* Heyne ex Roth (Moraceae) Umbar

Medicine

(*i*) Fruit latex is applied once daily inside mouth to cure mouth ulcer till cure.

(*ii*) Latex obtained after cutting leaves is applied daily on tumours till cure.

105. *Ficus religiosa* L. (Moraceae) Pimpal

Medicine

(*i*) A spoonful stem bark powder and a pinch of catechu (*Acacia catechu* Willd.) is dissolved in water. It is advised for 4-5 days to regularise menstruation.

(*ii*) Leaves are used as 'bidi' and smoked to treat migraine.

(*iii*) Leaves are extracted. About 25 ml of it is administered daily for few days to treat snake-bite.

(*iv*) 'Aag pain,' this hymn is written on five leaves at morning. This helps to cure all diseases.

106. *Gomphrena globosa* L. (Amaranthaceae) Supari

Medicine

(*i*) Root pieces are mixed in tea decoction and boiled. Two spoons of it is daily drunk at morning and evening on fever till cure.

107. *Grewia damine* Gaertn. (Tiliaceae) Mukahaman

Medicine

(*i*) A leafy branch is selected. It is defoliated. The stem is rubbed in water. This paste is wrapped in a betel leaf (*Piper betle* L.). It is consumed for daily three days to treat asthma.

108. *Hardwickia binata* Roxb. (Caesalpiniaceae) Anjan

Ethnoveterinary Medicine

(*i*) Leaves are fed to milching animals to increase milk yield.

Miscellaneous Uses

(*i*) Wood is used for agricultural implements and tools.

109. *Helicteres isora* L. (Sterculiaceae) Mural, Bahalsheng

Medicine

(*i*) Paste of fresh leaves and green fruits is applied daily on eczema till cure.

Ethnoveterinary Medicine

(*i*) About thirty fruits are crushed in lemon juice. [*Citrus aurantifolia* (Christm.)Sw.] It is administered to cows and buffaloes to treat dysentery.

110. *Hibiscus caesius* Garcke (Malvaceae) Ambadi

Medicine

(*i*) Leaves are cooked. Paste is prepared and is applied to injuries caused due to thorns.

111. *Hibiscus rosa-sinensis* L. (Malvaceae) Jaswand

Medicine

(*i*) Staminal columns of flowers are crushed in coconut oil (*Cocos nucifera* L.). It is applied on head as hair tonic.

(*ii*) Leaf juice, about half cup, is administered to cure dysentery for three to four days.

(*iii*) Leaves are chewed three to four times a day to cure mouth ulcer till cure.

112. *Holarrhena pubescens* (Buch.-Ham.)Wall. ex G.Don (Apocynaceae) Dudhi-indarjav

Medicine

(*i*) Seed powder, about spoonful, is consumed at morning and evening to cure colic for 5-6 days.

113. *Hygrophila schulli* (Buch.-Ham.)M.R. and S. M. Almeida. (Acanthaceae) Talimkhana

Medicine

(*i*) Seeds are extracted in milk. One spoonful of it is advised before meals for a fortnight as aphrodisiac.

114. *Jatropha gossypifolia* L. (Euphorbiaceae) Chandrajyoti

Medicine

(*i*) Young stem pieces are used to brush tooth at morning and night to increase strength of gums.

115. *Justicia adhatoda* Medic. (Acanthaceae) Adulsa

Medicine

(*i*) Decoction of leaves is prepared. A spoonful of it is advised for three times daily to cure cough till cure.

116. *Kedrostis rostrata* (Rottl.) Cogn. (Cucurbitaceae) Mirchikand

Medicine

(*i*) Tuber extract is mixed into cow-urine. It is applied daily on wounds and piles till cure.

117. *Kirganelia reticulata* (Poir.) Baill. (Euphorbiaceae) Pithori

Medicine

(*i*) Half cup of leaf juice mixed with equal curd at morning before tea is advised to treat piles till cure.

(*ii*) A cup of leaf juice alongwith lump sugar (*Saccharum officinarum* L.) is administered at early morning for seven days to treat impotency.

118. *Lagascea mollis* Cav. (Asteraceae) Pandhara-Purush, Safed-matha

Medicine

(*i*) Leaf juice (two spoons) is mixed in a cup of curd. It is consumed daily for seven days to treat kidney-stone.

119. *Langenaria siceraria* (Molina) Standl. (Cucurbitaceae) God-bhopla, Dudhi-bhopla

Medicine

(*i*) Leaf or fruit juice (about 15 ml) is taken orally at morning for seven days to cure jaundice.

(*ii*) Leaves are extracted. Half cup of extract and equal quantity of water is drunk daily at morning before food for three days to treat piles.

120. *Launaea procumbens* (Roxb.) Ramayya and Rajgopal (Asteraceae) Pathri

Medicine

(*i*) Leaves are cooked. Cooked leaves are then used as vegetable especially at night. This is advised to treat kidney-stone till cure.

(*ii*) Root extract is applied daily on piles till cure.

121. *Lavandula bipinnata* (L.) Ktze. (Lamiaceae) Ran-bhang

Medicine

(*i*) Leaves are dried and burnt. The smoke is inhalved to get rid of fever and cough.

122. *Lens culinaris* Medik. (Fabaceae) Masur

Medicine

(*i*) The pulse is crushed in coconut oil (*Cocos nucifera* L.) This paste is applied on hands and legs to cure rheumatism. It is advised till cure.

123. *Lepidagathis cuspidata* Nees (Acanthaceae) Bendya-ghas

Medicine

(*i*) Leaf powder is applied on wounds or scabies for three days.

124. *Leucas cephalotes* (Roxb. ex Roth) Spreng. (Lamiaceae) Dholkamni

Medicine

(*i*) Few pieces of dry roots are wrapped in a betel leaf (*Piper betle* L.). It is consumed at morning and evening to cure fever and pneumonia.

125. *Limonia acidissima* L. (Rutaceae) Kaith

Medicine

(*i*) Pulp of ripe fruit is homogenised with Turti (Alum.) Pellets are prepared. One pellet twice daily for seven days is consumed to treat piles.

126. *Luffa acutangula* (L.) Roxb.var. *amara* (Roxb.) Clarke (Cucurbitaceae) Kadu-dodki

Medicine

(*i*) Fruits are crushed and mixed with jaggery. Pellets are prepared. Two pellets are advised thrice a day for four to five days after dog-bite.

127. *Lycopersicum esculentum* Mill. (Solanaceae) Tamata

Medicine

(*i*) Unripe fruits are cut. Paste is prepared and applied for seven days to remove black spots on skin.

(*ii*) Consumption of ripe tomatoes for fifteen days or more is advised to recover from anemic condition.

(*iii*) Slices of ripe fruits are placed on black facial spots. This helps remove them if treated for seven days.

128. *Madhuca longifolia* (Koen.) Macbr. var. *latifolia* (Roxb.) Chevalier (Sapotaceae) Mahu

Medicine

(*i*) Raw fruits are rubbed in water. Two to three drops of this extract is poured into nose of a unconscious person.

(*ii*) Seeds are rubbed in water. Few drops are advised orally for a child to treat flatulence. It is applied to treat redness of eyes.

129. *Maerua oblogifolia* (Forsk.) A. Rich. (Capparidaceae) Dhor-gawat

Miscellaneous Uses

(*i*) It is used as fodder for cattle.

130. *Mangifera indica* L. (Anacardiaceae) Amba

Medicine

(*i*) Latex obtained after cutting leaves is applied on scorpion-sting.

(*ii*) Bark extract is applied daily on body to treat jaundice. It is then wipped out. This is followed till cure.

(*iii*) Raw fruits are crushed. The extract, one spoonful, is drunk for twice a day to treat dysentery.

131. *Maytenus emarginata* (Willd.) Ding Hou (Celastraceae) Henkal

Medicine

(*i*) Young leaves are chewed to treat tooth-ache.

(*ii*) Root bark is boiled in water. This decoction, five to ten ml, for two to three days is drunk to treat piles.

(*iii*) Leaf paste is heated and mixed in coconut oil (*Cocos nucifera* L.). It is applied daily on wounds and scabies till cure.

(*iv*) Leaf paste is dissolved in water. It is sieved through cloth. This infusion, about two to three spoons, is administered for three days to a woman suffering from leucorrhoea.

132. *Melia azedarach* L. (Meliaceae) Bajara-limb

Medicine

(*i*) Paste is applied on dog-bite for 31 days.

133. *Mentha spicata* L. (Lamiaceae) Pudina

Medicine

(*i*) Fresh leaves are consumed to treat cough.

(*ii*) A spoonful leaf juice mixed in equal cow-ghee is administered once or twice to control stomach-pains.

134. *Merremia dissecta* (Jacq.) Hall.*f.* (Convolvulaceae) Afumari

Medicine

(*i*) Few leaves are cut into pieces. These are added in boiling cow-ghee. This ghee, 2-3 drops, is poured into nose as a remedy against migraine. It is advised daily once for three to four days.

(*ii*) A spoonful of leaf powder is consumed twice a day to treat diabetes for a week.

135. *Michelia champaca* L. (Magnoliaceae) Sonchafa

Miscellaneous Uses

(*i*) Flowers are used while worshipping and praying gods by Hindus.

136. *Mimosa hamata* Willd. (Mimosaceae) Arati

Medicine

(*i*) Roots are boiled in water. This decoction is advised about, half cup at morning and evening, to treat rheumatism till cure.

137. *Momordica charantia* L. (Cucurbitaceae) Karle

Medicine

(*i*) Fruits are dried and powdered. A spoonful of it in honey is administered for seven days at morning to control diabetes.

(*ii*) Fruit juice, about 50 ml, is drunk at morning for four to five days to cure piles.

(*iii*) Seeds are extracted. A spoonful of it is consumed to cure colic. It is advised for five to six days.

138. *Momordica dioica* Roxb. ex Willd. (Cucurbitaceae) Katurle

Medicine

(*i*) Tuber is crushed and paste is prepared. About 5 gm of it is consumed at morning before consumption of food. It is consumed by a woman for seven days to conceive. It is also useful to regularise menstruation.

Ethnoveterinary Medicine

(*i*) Roots are extracted in water. This extract, about 2-3 glasses, is administered orally to a cow to stop bleeding after delivery.

139. *Morinda pubescens* J.E.Sm. (Rubiaceae) Bardoi

Medicine

(*i*) Bark extract, about five ml, is administered daily once for seven days to treat jaundice.

(*ii*) Ripe fruits are consumed to purify blood. Two fruits per day are advised for a fortnight.

(*iii*) Seeds of this plant are crushed in oil. This extract is applied daily at night on knees to check rheumatism.

140. *Moringa oleifera* Lamk. (Moringaceae) Shevga

Medicine

(*i*) Leaf extract, few drops, are added into nose as a remedy against head-ache. It is advised only once a day till cure.

141. *Murraya koenigii* (L.) Spreng. (Rutaceae) Kadhi-patta

Medicine

(*i*) Consumption of leaves in diet is beneficial to keep heart healthy.

142. *Musa paradisiaca* L. (Musaceae) Kela

Medicine

(*i*) Pulp of raw fruit is daily applied on eczema and wrapped by cloth till cure.

(*ii*) Young unripe fruit is consumed at morning to cure piles till cure.

143. *Nicotiana tabacum* L. (Solanaceae) Tambakhu

Medicine

(*i*) Five to six gm of leaf powder and 100 gm of sugar (*Saccharam officinarum* L.) are mixed thoroughly. It is consumed once daily to control dysentery till cure.

144. *Ocimum americanum* L. (Lamiaceae) Ran-tulsi, Jangali- tulsi

Medicine

(*i*) Few leaves are wrapped in a betel leaf (*Piper betle* L.). It is consumed daily once to check cough.

(*ii*) Few leaves and seeds are crushed and paste is prepared. It is applied daily on wounds till cure.

(*iii*) Roots are first dried. These are powdered and paste is prepared in water. It is consumed, about 3-4 gm, daily once for seven days to control fever.

145. *Ocimum basilicum* L. (Lamiaceae) Sabja

Medicine

(*i*) Leaf juice of this plant and those of Tulsi (*Ocimum tenuiflorum* L.) alongwith some sugar (*Saccharum officinarum* L.) and cow-ghee are homogenised. This mixture is applied on head. Few drops are also added into nose at the same time to treat severe head-ache.

(*ii*) Leaves of 'Sabja', a leaf of mango (*Mangifera indica* L.), a leaf of 'Jambhul' (*Syzigium cumini* (L.) Skeels and few leaves of 'Babhul' (*Acacia nilotica* L.) Del. subsp. *indica* (Bth.) Brenan are dried and powdered together. This powder is used to clean tooth daily for three days. This helps to remove foul smell of mouth and prevents pus formation of gums and tooth.

146. *Ocimum tenuiflorum* L. (Lamiaceae) Tulsi

Medicine

(*i*) Leaf juice, two to three drops, is poured into ears to treat ear-ache till cure.

(*ii*) Leaf juice and 'black matter' of dry battery cell are homogenised. This is applied to treat eczema for two to three days.

(*iii*) Leaf juice, about spoonful twice daily, is drunk to cure fever till cure.

(*iv*) Decoction of four leaves of this species and four leaf-lets of Neem (*Azadirachta indica* A. Juss.) is prepared in cow-urine. This decoction is drunk at morning till cure. It is beneficial in heart complaints, ulcer and rheumatic pains.

(*v*) Few leaves are wrapped in a betel leaf (*Piper betle* L.). It is consumed at morning and evening to treat bile.

(*vi*) Ten leaves are boiled in the decoction of tea. This tea is drunk twice daily as a remedy against cough.

147. *Opuntia elatior* Mill. (Cactaceae) Nivdunga

Medicine

(*i*) Fruits are heated and powdered. A spoon of it is consumed at morning to cure rheumatism till cure.

(*ii*) Flower juice is taken orally daily once to cure whooping cough.

(*iii*) Bark is heated and extracted. It is applied around ears to cure ear-ache.

148. *Pennisetum americanum* L. (Poaceae) Bajari

Medicine

(*i*) Flour paste is prepared in water. It is applied on blisters daily.

149. *Pergularia daemia* (Forsk.) Chiov. (Asclepiadaceae) Utranda

Medicine

(*i*) Leaf extract is mixed with coconut oil (*Cocos nucifera* L.) in 1:3 ratio. A half cup of it is taken orally once daily to cure dysentery.

(*ii*) Leaf extract is drunk, three to four spoons a day for a week, to check acidity.

(*iii*) Leaf extract, few drops, is poured into ears to avoid pus formation.

(*iv*) About half spoon of leaf extract is advised for children at morning and evening for deworming intestinal worms.

(*v*) Leaf paste is applied on head at morning and evening to grow hair on bald till hair appears.

(*vi*) Leaves are crushed alongwith common salt. This paste is applied daily once on tumour developed after thorn injury till cure.

Ethnoveterinary Medicine

(*i*) Leaf extract is applied on tumours, blisters and leg-ache in case of bullocks.

150. *Phyllanthus airy-shawi* Burnel and Roux (Euphorbiaceae) Futani-gawat

Medicine

(*i*) Leaf juice, about one spoonful, is drunk at morning and evening to treat amoebic dysentery for a week.

151. *Phyllanthus amarus* Schumach and Thonn. (Euphorbiaceae) Bhui-awala

Medicine

(*i*) Entire plants are extracted. About 10-15 ml of it is drunk at morning for three to five days to cure jaundice.

152. *Pithecellobium dulce* (Roxb.) Benth. (Mimosaceae) Vilayati-chinch

Miscellaneous Uses

(*i*) It is used as fodder.

153. *Plumbago zeylanica* L. (Plumbaginaceae) Safed-chitrang

Medicine

(*i*) Root powder is consumed twice a day for a week to purify blood.

154. *Psidium guajava* L. (Myrtaceae) Peru, Jam

Medicine

(*i*) About five to six leaves are chewed and saliva is swallowed to control acidity. This is advised for few days.

(*ii*) Extract of leaves, two to three spoons, is consumed for few days to purify blood. It is also beneficial in controlling diarrhoea.

155. *Punica granatum* L. (Punicaceae) Dalimb

Medicine

(*i*) Fruit walls are extracted in water. It is drunk at morning and evening to control cough till cure.

(*ii*) Fruit walls are dried and powdered. Some algal filaments (in 1:1ratio by weight) are homogenised with this powder. It is heated and then applied for piles. Consumption of fruits of *Momordica charantia* L. and candy (*Saccharum officinarum* L.) is advised during the period of treatment.

156. *Raphanus sativus* L. (Brassicaceae) Mula

Medicine

(*i*) Root extract is prepared in water. About one glass of it is drunk for three to four days at morning to cure piles.

157. *Ricinus communis* L. (Euphorbiaceae) Erandi

Medicine

(*i*) A spoonful of seed oil is drunk as purgative.

(*ii*) Leaf extract is applied on gums to treat tooth ache.

(*iii*) Leaf extract and seed oil are together applied on knees at night to treat rheumatic pains.

(*iv*) Petiole is burnt. The fumes are inhalved once a day to control paralysis till cure.

(*v*) Leaf paste and root paste, about a spoon each, are mixed in a cup of milk. It is consumed at twice a day to treat jaundice till cure.

(*vi*) Flowers and sugar-candy (*Saccharum officinarum* L.)are mixed thoroughly. Pellets are prepared: A pellet at morning and evening is advised to treat jaundice till cure.

(*vii*) Root pieces are kept in between tooth to cure tooth ache.

(*viii*) Flowers are garlanded around the neck of a patient to treat jaundice till cure.

(*ix*) Leaves are boiled in water. Few drops are mixed in few drops of sesame oil (*Sesamum orientale* L.) to cure ear-ache.

(*x*) Flower juice (10 ml) mixed in milk (half cup). It is drunk for three days at morning to cure jaundice.

158. *Rottboellia cochinchinensis* (Lour.) W.D. Clayton (Poaceae) Pendhi

Miscellaneous Uses

(*i*) It is used as fodder for cows and buffaloes.

159. *Saccharum officinarum* L. (Poaceae) Oos

Medicine

(*i*) Fresh sugarcane juice is drunk daily to treat jaundice till cure.

160. *Salvadora persica* L. (Salvadoraceae) Pillu-pala

Medicine

(*i*) Few leaves are regularly consumed for 21 days to treat tuberculosis.

Ethnoveterinary Medicine

(*i*) Leaf extract and whey are mixed in 1:1 ratio. About 3-4 cups of it is administered daily to bullocks to treat flatulence.

161. *Selaginella ciliaris* (Retz.) Spring (Selaginellaceae) Sanjivani

Medicine

(*i*) Entire plants are dried and powdered. It is mixed with candy (*Saccharum officinarum* L.). One spoonful of it is mixed in curd which is then consumed. This is followed to cure piles and also to regularise menstrual cycle.

162. *Semecarpus anacardium* L.*f.* (Anacardiaceae) Biba, Bani, Bhilawa

Medicine

(*i*) Seed is heated and the oil expressed is applied on injury caused due to thorns.

(*ii*) Seeds ash is mixed into coconut oil (*Cocos nucifera* L.). It is applied on eczema.

(*iii*) Seed oil is mixed in milk. It is, one spoonful daily for three days, is drunk to control diarrhoea. This paste is also useful in case of tetanus.

(*iv*) Seeds paste is applied on wounds.

Ethnoveterinary Medicine

(*i*) Crushed seeds are homogenised with cow-dung. It is applied on daily around neck to check throat swelling in case of pet animals till cure.

Miscellaneous Uses

(*i*) Seeds are burnt and are kept in the holes where in a snake has entered. The snake inside thereby dies.

163. *Sesamum orientale* L. (Pedaliaceae) Til

Medicine

(*i*) Leaf juice, two to three drops, is poured into ears to prevent infection.

(*ii*) Few drops of leaf juice are added in an ear of opposite side of aching tooth. This reduces ear-ache.

164. *Sesbania bispinosa* (Jacq.) Steud. ex Wight (Fabaceae) Kath-shevari

Medicine

(*i*) Flowers are dried and powdered. The powder is smelt to stop nose bleeding.

(*ii*) Leaf juice and powder of inner bark both are mixed homogenously. It is added in a cup of whey. A cup of it is daily drunk for a mouth to control excessive bleeding during menstruation.

(*iii*) Bark is soaked in water. Infusion is obtained. Candy (*Saccharum officinarum* L.) is added in it and cup of it is administered for three days to check leucorrhoea.

(*iv*) Extract of bark, about 10-15 ml thrice a day, is advised to check blood dysentery.

165. *Sesbania grandiflora* (L.) Poir. (Fabaceae) Hadga

Medicine

(*i*) Cooked flowers are used as vegetable to improve digestion.

166. *Sesbania sesban* (L.) Merr. (Fabaceae) Kath-shevri, Shewari, Uskatyari

Medicine

(*i*) Root paste, about 4-5 gm, is mixed in a cup of milk. It is consumed twice daily for 5-6 days to treat typhoid. Non-vegetarian diet is avoided during the treatment.

(*ii*) Roots are chewed daily once to treat mouth ulcer.

(*iii*) Bark of stem is extracted in water. This extract, about 20 ml, is drunk thrice daily to treat blood dysentery till cure. It is also advised to treat leucorrhoea.

167. *Solanum anguivi* Lamk. (Solanaceae) Pandhare-dorle

Ethnoveterinary Medicine

(*i*) Leaf extract is applied on eyes of a bullock suffering from eye-ache.

168. *Solanum nigrum* L. (Solanaceae) Kamoni

Medicine

(*i*) Leaf juice is used to massage body for checking rheumatic pains. This is applied for two to three days.

(*ii*) Leaf juice is mixed in 1:2 ratio in cow-milk. Half cup of it is drunk daily once to cure jaundice. This is followed till cure.

169. *Solanum virginianum* L. (Solanaceae) Bhui-ringani

Medicine

(*i*) Seeds are wrapped in a betel leaf (*Piper betle* L.). It is consumed to cure intestinal torsion. Food before and after the dose is avoided for some hours.

(*ii*) Fruit juice mixed with curd (in 1:1 ratio) is applied on gums to treat tooth-ache.

(*iii*) Roots are boiled in water. This decoction, about 20 ml, is drunk two to three time a day as a remedy against stomach-ache.

(*iv*) Fruit is cut and common salt is filled in it. It is then heated and tied on injury caused due to thorns. This is followed for three to four days.

(*v*) Root paste is prepared in milk. Three to four gm of it wrapped in a betel leaf (*Piper betle* L.) and consumed daily to cure 'Nagin' disease till cure.

(*vi*) Paste of fresh roots is applied on eczema daily till cure.

(*vii*) Roots are crushed. Decoction is prepared. A spoonful of it a day is advised as a remedy against tuberculosis till cure.

Ethnoveterinary Medicine

(*i*) Fruit juice is applied on bullock's eyes to better eye sight. It is followed for three to four days.

170. *Sorghum bicolor* Moench (Poaceae) Jawar, Jwari

Medicine

(*i*) Grains are burnt. Their ash is mixed in common salt in 1:1 ratio. It is used as tooth powder twice daily for 5-6 days to remove foul smell of mouth.

171. *Soymida febrifuga* (Roxb.) A. Juss. (Meliaceae) Rangat-roli

Medicine

(*i*) Stem bark is powdered. About 5 gm is dissolved in a half cup of cow-milk. It is drunk for 15 days to treat spermatorrhoea.

(*ii*) Bark powder, about one spoonful, is advised at morning and evening to purify blood for 7-8 days.

172. *Spermadictyon suaveolens* Roxb. (Rubiaceae) Padri

Ethnoveterinary Medicine

(*i*) A root extract, about 150 gm, mixed in flour is fed to cows and bullocks to check blood dysentery.

173. *Sphaeranthus senegalensis* DC. (Asteraceae) Gorakhmundi

Medicine

(*i*) Flowers (3-4 heads) are consumed to increase appetite daily for 4-5 days.

174. *Spilanthes calva* DC. (Asteraceae) Akkal-kadha

Medicine

(*i*) Flowers are crushed in honey and administered at morning to children in case of stammering.

175. *Sporobolus coromandelianus* (Retz.) Kunth (Poaceae) Kuran

Miscellaneous Uses

(*i*) It is used as fodder for cattle.

176. *Syzigium cumini* (L.) Skeels (Myrtaceae) Jambhul

Medicine

(*i*) Decoated seeds are crushed in cow milk. A spoonful of it is consumed at morning and evening to check diabetes.

(*ii*) Bark is boiled in water. A cup of it is drunk at morning and evening to better intestinal pains.

(*iii*) Dry fruit powder, one spoonful, is consumed at morning and evening for seven days to cure diabetes.

(*iv*) (a) Leaf powder is applied on gums to treat tooth ache. (b) It is also applied inside mouth to check mouth ulcer.

(*v*) Ripe fruits are consumed at morning before food to control blood pressure.

177. *Tagetes erecta* L. (Asteraceae) Zendu

Medicine

(*i*) Leaf juice, about 5 to 10 ml, is drunk to treat intestinal worms.

(*ii*) Leaf juice is applied inside ears to check to cure ear-ache.

(*iii*) Leaf juice is applied daily onto abdomen of a child to deworm intestinal worms till cure.

178. *Tamarindus indica* L. (Caesalpiniaceae) Chinch

Medicine

(*i*) Seeds are bolied in water. It is applied on scorpion sting.

Ethnoveterinary Medicine

(*i*) Leaves are boiled in water. These leaves are used to deworm legs of pet animal.

179. *Tectona grandis* L.*f.* (Verbenaceae) Sagwan

Medicine

(*i*) Seeds are crushed and powdered. It is, 5 gm daily at morning, consumed for 8-10 days to cure kidney stone.

(*ii*) A seed paste is applied around scrotum-sacs to treat sunstroke.

(*iii*) Dry leaves are burnt and the ash is applied daily on skin to treat skin allergy till cure.

(*iv*) Seeds extract (about 15 ml) mixed in water is drunk once daily to cure kidney-stone till cure.

180. *Tephrosia purpurea* (L.) Pers. (Fabaceae) Diwali

Medicine

(*i*) Stem bark is crushed in water. About 10-15ml is taken orally at morning and night as a remedy against kidney-stone till cure.

(*ii*) Root powder, about 5 gm daily, is consumed for seven days at morning and evening to treat kidney-stone.

181. *Terminalia arjuna* L. (Combretaceae) Arjun-sadada

Medicine

(*i*) Powder of bark mixed with coconut oil (*Cocos nucifera* L.) is applied daily to cure injury till cure.

(*ii*) Seeds ash is mixed in coconut oil (*Cocos nucifera* L.) or bark powder is mixed in coconut oil and is used to avoid infection of injuries.

(*iii*) Leaf paste is applied on bone fracture and cramps till cure.

182. *Terminalia bellirica* (Gaertn.) Roxb. (Combretaceae) Behda

Medicine

(*i*) Fruits and inner bark are extracted in water. About 10-15 ml of it is drunk at morning to treat whirling.

183. *Terminalia chebula* Retz. (Combretaceae) Hirda

Medicine

(*i*) Leaf paste is applied daily on scabies till cure.

(*ii*) Seed powder is used as tooth paste to clear tooth-ache.

184. *Thespesia populnea* (L.) Soland. ex Corr. (Malvaceae) Paras-pimpal

Medicine

(*i*) Fruits are extracted alongwith jaggery. It is sieved through cloth and pellets are prepared. One pellet a day for 15 days is advised to a man and woman to treat impotency.

(*ii*) One pellet a day is advised to woman for a fortnight to treat impotency.

Ethnoveterinary Medicine

(*i*) The aforesaid pellets are fed mixed in fodder of pet animals to arouse sex. The dose is, however, larger. About 50 gm of the said recipe is advised for 15–21 days.

185. *Tinospora cordifolia* (Willd.) Miers. (Menispermaceae) Gulwel

Medicine

(*i*) Leaves, about 10 ml, are drunk after meal daily to cure fever till cure.

(*ii*) Root extracts and decoction of tea in equal ratio is drunk at morning and evening to cure jaundice, whooping cough and asthma till cure. About 3-4 spoons of it is advised.

(*iii*) Root juice, about two spoons, is drunk thrice daily for three days to treat typhoid.

(*iv*) Leaves are warmed and then juice is obtained. It is mixed in cow-milk in 1:1 ratio. A cup of it is drunk at morning for three days to cure jaundice.

(*v*) Leaf juice is advised to treat diabetes. It is drunk, about three spoons daily, till cure.

186. *Trianthema portulacastrum* L. (Molluginaceae) Pandhara-vasu

Medicine

(*i*) Root extract, about 15 ml twice daily, is administered to treat jaundice.

187. *Tribulus lanuginosus* L. (Zygophyllaceae) Sarnati

Medicine

(*i*) Leaf juice is mixed with milk in equal quantity. A spoonful of it is administered thrice daily to regularise menstruation till cure.

188. *Tribulus terrestris* L. (Zygophyllaceae) Gokharu

Medicine

(*i*) Fruit thorns are ground alongwith candy (*Saccharum officinarum* L.). It is mixed in a cup of water. It is consumed twice a day to treat gastritis for three days.

(*ii*) Fruit powder is mixed in milk and sugar (*Saccharum officinarum* L.). Paste is prepared. This paste is applied on rheumatism and for body pains.

189. *Trichosanthes cucumerina* L. (Cucurbitaceae) Sonpadol

Medicine

(*i*) Seeds are wrapped in a betel leaf (*Piper betle* L.). It is chewed to daily to cure jaundice.

(*ii*) Fruits are crushed and boiled. The juice one teaspoon thrice daily is advised to treat asthma till cure.

190. *Trichosanthes tricuspidata* Lour. (Cucurbitaceae) Gavalan

Medicine

(*i*) Decoction is prepared using two seeds and rhizome of

ginger (*Zingiber officinale* Rosc.). Half cup of it twice daily is advised to treat stomach-ache till cure.

191. *Tridax procumbens* L. (Asteraceae) Tantani, Tongalmodi, Ghavpala

Medicine

(*i*) Fine paste of leaves is applied at morning on the infected portion of mouth of children daily.

(*ii*) Two to three drops of leaf juice is added into ears to cure tooth ache.

(*iii*) Leaf paste is applied on wounds till cure.

(*iv*) Flowers alongwith camphor (*Ocimum kilimandscharium* Guercke) are extracted. This extract, 1-2 drops, is applied around ears to stop ear-ache.

(*v*) Leaf juice, about 10-15 ml, is mixed in a cup of milk. It is drunk once daily to check dysentery for 2-3 days.

(*vi*) Leaf extract is applied on fresh cuts or injuries to prevent infection.

192. *Trigonella foenum-graecum* L. (Fabaceae) Methi

Medicine

(*i*) A spoonful seed powder is consumed at morning and evening to cure rheumatism till cure.

(*ii*) Seeds of 'Methi', wawding (*Embelia tsjeriam-cottam* DC.), 'Nirgudi' (*Vitex negundo* L.), and candy (*Saccharum officinarum* L.) are powdered finely. About 5 gm is homogenised in cow-ghee and consumed daily to treat rheumatism till cure.

193. *Triticum aestivum* L. (Poaceae) Gahu

Medicine

(*i*) Leaves are obtained from young growing seedlings. Their extract, about 25 ml, is drunk to treat stomach-ache and constipation.

194. *Triumfetta rhomboidea* Jacq. (Tiliaceae) Landaga

Medicine

(*i*) Leaf extract is applied on wounds daily to deworm them. This is followed for 2-3 days.

195. *Tylophora dalzellii* Hook. *f.* (Asclepiadaceae) Kali-kavali

Medicine

(*i*) Roots are extracted in water. This extracts, about 10 ml, is taken orally for 2 to 3 days at morning and evening to treat kidney-stone.

196. *Typha domigensis* Pers. (Typhaceae) Pankanis

Medicine

(*i*) Fresh inflorescence is crushed. This paste is applied daily for eczema till cure.

197. *Verbascum chinense* (L.) Sant. (Scrophulariaceae) Kar

Medicine

(*i*) Leaf paste is applied daily on eczema till cure.

198. *Vernonia anthelmintica* (L.) Willd. (Asteraceae) Kadu-jeera

Medicine

(*i*) Seeds are crushed and then boiled in water. This decoction, about 10-15 ml, is drunk for three days at morning to reduce body fever.

199. *Vernonia cinerea* (L.) Less (Asteraceae) Sahdevi

Miscellaneous Uses

(*i*) Leaves are used as fodder for cattle.

200. *Ventilago denticulata* Willd. (Rhamnaceae) Sakhal-wel

Medicine

(*i*) Bark is boiled in water. This decoction is applied on gums daily to check the tooth-ache till cure.

(*ii*) The tender part of stem is extracted. It is, 2 to 3 drops, added into ears to get relief from head-ache.

Ethnoveterinary Medicine

(*i*) Leaf extract, about a litre, is administered orally to bullocks to check dysentery. It is advised daily once for 3-4 days.

201. *Vitex negundo* L. (Verbenaceae) Nirgudi

Medicine

(*i*) Powder of dried roots and leaves in equal quantity is mixed. About 5 gm of it daily for forty days is consumed to cure diabetes and ulcer.

(*ii*) Leaves are boiled in water. This water is used for bath daily to treat rheumatism till cure.

(*iii*) Leaves are heated and tied onto legs to treat rheumatism for a week.

202. *Withania somnifera* (L.) Dunal (Solanaceae) Ashwagandha, Askand

Medicine

(*i*) Tubers are dried and powdered. This powder is, 15 gm per day, is consumed for seven days to treat waist-ache and to also as aphrodisiac.

(*ii*) Roots are powdered. This powder, 2-3 gm, is added while preparing sweet-balls. Each of it is consumed daily to treat rheumatism.

(*iii*) Root powder is mixed in water and 5-10 ml of it is administered daily after delivery to overcome weakness. It is advised till cure.

(*iv*) Ash obtained after burning egg-shell, powder of roots of this plant and fruits of 'Ritha' (*Sapindus emarginatus* Vahl) are mixed in equal quantity. This mixture is applied inside ears to stop pus formation. Tea drinking is avoided during the period of treatment.

(*v*) Leaves of this plant and leaves of castor (*Ricinus communis* L.) are extracted together. Half cup of it is drunk daily to cure jaundice till cure.

(*vi*) Root extract, about 20-25 ml, is taken orally at night for 15 days to increase body strength.

203. *Xanthium indicum* L. (Asteraceae) Bedyache tree, Nakeshwar, Gokharu

Medicine

(*i*) One dried leaf is powdered alongwith fruits of cardamom

[*Elettaria cardamomum* (L.) Maton]. This is consumed at early morning before food. This is advised for seven days to treat urinary stone. It is also advised for sun-stroke.

(*ii*) Seed powder, about 2 gm, is mixed in a cup of cow-milk. It is advised for seven days to treat jaundice.

(*iii*) Dry leaf powder mixed with cow-milk and lump sugar (*Saccharum officinarum* L.) is drunk. One spoonful of it is advised at morning to cure rheumatism.

(*iv*) Leaf juice is mixed with lime. It is applied on injury caused due to thorns.

204. *Ziziphus mauritiana* Lamk. (Rhamnaceae) Bor

Medicine

(*i*) Leaf extract and equal quantity of cow-ghee are homogenised. One spoonful of it daily for four days is drunk to cure piles.

(*ii*) Leaves are heated and powdered. It is mixed in coconut oil (*Cocos nucifera* L.). This paste is applied on wounds and scabies till cure.

(*iii*) Leaf paste is mixed in sheep-dung. It is heated in a pot. It is then applied near anus to treat piles till cure.

(*iv*) Extract obtained from young leaves is advised at early the morning to reduce body heat.

205. *Ziziphus nummularia* (Burm.*f.*) Wight and Arn. (Rhamnaceae) Yer-bhol

Medicine

(*i*) Roots are powdered. About 2 gm of it is mixed in decoction of tea. Milk is avoided. A cup of this decoction is advised at morning to treat fever for three to four days.

PART – III
DISCUSSION AND CONCLUSIONS

Traditional knowledge refers to the knowledge, practices and innovations of indigenous communities. It is transmitted orally from generation to generation and is developed from experience gained over the centuries and adapted to the local culture. It is generally in the form of folklore, cultural values, rituals, stories, even songs or proverbs, etc. A large proportion of population especially of the developing countries like India still relies on traditional healthcare and for other miscellaneous needs.

Mankind has been using biota since his emergence on this planet. He lived close to the nature and hence acquired knowledge about it. He was equipped with deep and discerning acquaintance with the flora in his surroundings. He gradually acquired knowledge of properties of several plant species. This knowledge became an integral part of his culture which handed down to the next generations. The tribal and rural ethnic communities have still retained it to a large measure. As the ethnic groups migrated from place to place in search of their livelihood, their folklore knowledge also became sometimes fragmented and travelled with them with additions or deletions.

There is substantial research work going on in our country to document traditional knowledge. Many research workers have been filliped. In India, Dr. S.K. Jain inspired scientists to undertake such records. As far as Maharashtra state is concerned, at the beginning, the western part of the state is botanised on this line by late Dr. V.D. Vartak and his associates. The ethnobotanical work has also extended in the parts of Vidarbha region of the state. However, Buldhana district till remained untouched. In the following, the results of our work are discussed in tehsils under study. The present investigation suggests that the people of the Buldhana district have a fairly and detailed knowledge regarding plants and their utilities whether wild or cultivated. The inhabitants have a wide knowledge of applications or uses of plants for different purposes including medicinal, food, fodder, fire-wood, timber, agricultural implements and tools etc. (Appendix I to X). The present attempt revealed that total of mainly 412 species belonging to 307 genera and 91 families are in vogue for different purposes. The dicotyledons are represented by 55 families, 259 genera and 355 species, whereas the monocotyledons are represented by 54 species. Total 58 species of crop weeds also find place in food, fodder or medicine. The different uses of plants are discussed herewith.

Majority of plants are useful as medicine for both human-beings and his pet animals. Majority of species are employed for various medicaments of the former, whereas 90 species are useful for the latter (Appendix: I). The inhabitants employ roots, other underground parts, stem, leaves, flowers, fruits, seeds, gum, latex, etc. They mostly find suitable recipes in the form of decoction, infusion, paste, powder, extract, juice, oil, ash, fume or their products in the form of sweet balls and pellets etc. They also mix the substances like milk, curd, whey, various edible oils, sugar, honey, jaggery, red lead, lime, flour, camphor, candy, ant-hill soil, animal ghee, urine of pet animals e.g. cow and goat. The plants parts are sometimes heated, cooked or raw. In few cases, entire plants are employed in preparing recipes. Apart from principal drug plants, about 74 other plants or their parts are used to supplement the drugs. (Appendix: II). Some drugs are not derived directly from the species growing in this area, but they are purchased from the local markets. All these are plant derived substances normally used for food or food adjuncts e.g. *Phoenix dactylifera, Areca catechu, Piper nigrum, Vitis vinifera, Syzygium aromaticum, Trachyspermum ammi, Nicotiana tabacum, Piper betle*, etc. In some cases, especially the local medicine-men utter certain hymns or 'mantras'. This is followed especially in cases where the sufferer and his medicine-men believe in supernatural powers of spirit. They associate some diseases/sufferings with the anger of the gods, goddesses or even ghosts. In few cases, it appeared that the medicine-man treats the patients psychologically while murmuring such hymns. This study indicated that the people have deep faith in local remedies. They treat various ailments such as migraine, head-ache, cold, cough, ear-ache, pus formation in ears, stomach ache, liver disorders, intestinal worms, acidity, diabetes, asthma, mouth-ulcer, tooth-ache, molar-ache, axima, itching, dandruff, suppurating tumours, dysentery, impotency, whitlow, sun-stroke, wounds, injuries, burns, spermatorrhoea, leucorrhoea, irregular menstrual cycles, complaints about conception, hepatitis, general body-ache, eye, knee-ache, rheumatism, bone fracture, urinary or kidney stone, fever, typhoid, malaria, accidental snake-bite, scorpion stings, dog-bite and few other human diseases.

Responses to our queries about the herbal remedies prevalent in this region are interesting. The people consider the medicines safe. No unusual symptoms develop due to them. They avoid certain medicines also during pregnancy and empty stomach. However, the younger generation appeared unwilling to learn traditional skills. The immediate preference of local remedies also appeared due to the lack of modern medicine, healthcare, poverty and high cost medicine.

Crop weeds are also curative for some ailments. Few are also used as leafy vegetables. The economic potential of such weedy species should be explored. Agriculture is the mainstay for sustenance and hence crop weeds are easier for their access.

To maintain oral hygeine, natural and artificial methods are used. Oral cavity acts as natural habitat for microorganisms. Tounge, tooth, saliva, mucous membrane lining, the lips, cheek and palate are ideal habitats for pathogens. Traditionally as in other parts of our country, the people of this area employ fruits for mouth wash or twigs as toothbrush (i.e. Datan or datun). In modern times, oral health is maintained by use of synthetic chemical products. These are also used by the rural people. These synthetic dentifrices have their own disadvantages. These may damage gum tissue or recession of the gums. The natural and traditional methods are said to be more beneficial. Active principles of stem brushes (datun) should be studied. This study will help recommend some harmless and useful drugs.

The body is delimited externally by the skin. Hair and nails are also its derivatives. The skin is pliable and protective covering. It regulates body temperature and also controls excessive water loss. It is important area of storage and synthesizer of many substances in the body. It receives a variety of sensations. This indicates that the skin is one of most complex structure. It has to perform many functions. It is easily exposed to various environments, dirt and pathogens. It is the first structure to be disturbed in accidental works. It is therefore a victim of many ailments like wounds, injuries, rashes, itches, dermatitis, skin infections, warts burns, dogs bite, insect bite, snake bite, scorpion sting, leprosy, etc. To treat all these skin affictions, a large numbers of treatments are known to the inhabitants of these three tehsils. The recipes employed are nurturing in nature, curative or even preventive. The plant species and their parts/products have their own active principles but these need to be examined on scientific basis.

A considerable number of species are used for the treatment of wound healings. Such species are available generally in close vicinity and therefore mostly fresh plants or their products are employed to treat them. It is worthwhile to gather knowledge on such species and derivatives with special reference to their mechanism of action.

Indian subcontinent is facing the menace of population explosion. National programmes have been initiated or in operation for family planning. Several personal enquiries to reveal particular contraceptives in this region mostly failed. The people generally hesitate to reveal their medicaments. Few complaints about spermatorrhoea, leucorrhoea, menstrual cycle, etc. are recorded from this area. It is also the case about urinary and kidney stones. Shyness is another factor for not involving into discussion.Generally; minor self-limiting ailments are traditionally treated within the family using remedies passed down through generations. There appeared certain instances when household remedies failed to cure, the people in area also sought for sophisticated professionals depending upon their financial status.

The respiratory disorder *viz.*, asthma is a very common health problem in this part of the district. The symptoms like tightening of small bronchial tubes, production of mucus making breathing difficult, having in chest and restlessness are inhabited by the sufferers themselves. After a long period of suffering, it is usually identified. People use few plant species to combat asthma. Probably, these plants contain very specific product which cures this disorder. The respiratory system actually stretches from the linning of the eyes and sinuses to the base of the lungs. These are generally exposed to organisms in air, dust or dirt. Remedies for these are also prevalent in these societies. The plants useful for these complaints generally soothe the mucous membranes and relieve inflammation of the related body part.

Amongst the rural communities, stomach complaints are commonly met with, for example flatulence, dysentery, diarrhoea, intestinal worms etc. Clean water facility is not generally available throughout the year. During monsoonal period, the complaints are more frequent. The inhabitants are also unable to diagnose correctly the diseases of digestive system. In such cases, the help of seniors as expert medicine-men is required.

Snake-bites, scorpion-strings or dog-bites are common especially in rural parts and the communities involved in agricultural works. It is said that although the number of species of snakes is light, only few, e.g. cobra, common kraits, viper, etc. are deadly poisonous to man. But the most common symptoms after snake, whether poisonous or non-poisonous, is sudden fear. On account of it, the victim mostly becomes semi-conscious with cold clammy skin, feeble pulse and rapid breathing. These are actually emotional symptoms. Sometimes, it produces psychological shock and even death. The experts for such complaints use the natural resources available in their surrounding. They believe in 'tantra' and 'mantra'. Such plant species which are commonly are employed are listed under Appendix: III.

Some patients in this area were observed under treatment for contact therapy. A part of a medicinal plant is made to touch the body of the sufferer. It is tied to the diseased part of body. Sometimes the organ of a patient is wrapped by the parts or the plant part is hung around waist or around neck. Observations are made on some of the diseases cured. The present investigators recorded about 20 plant species used to avoid various diseases. Such practices are also common in other Indian communities (*cf.* Peter, 1996; Punjani, 2002).

The present investigators also noted that species employed especially by women to treat that diseases like excessive bleeding during menstruation, irregular menstruation, various urinary complaints, impotency, pre-and post-delivery complaints, etc. Similarly, men from this area employ some plant species in treating sperm count or production, impotency, sterility, sexual desire, etc.

Man, since his appearance on this planet has a desire to look beautiful. He also, therefore, tried to obtain useful products from nature. These are several plant species which are beneficial in skincare. The various skin diseases, cracking and wrinkling of skin, discoloration of skin and aging are being treated by plant-derived products. Fruits of *Sapindus emarginatus, Emblica officinalis,* root tubers of *Cyperus rotundus,* flowers of *Hibiscus rosa-sinensis* and leaf pulp of *Aloe vera* do not need now further recognition. These are commonly employed by the elites. All these and such others have been known to our ancestors. The present investigator noted some such species.

Kidney stones are one of the most painful of the urologic disorders. It is also one of the most common complaints. A kidney stone is a hard mass developed from crystals and build up on the inner surface of the kidney. It is mostly composed of calcium in combination with either oxalate or phosphate. These chemicals are part of our diet. The stones also develop in the urinary tract due to infection. It appeared from the observations and discussions with inhabitants of this area that the prevalence of kidney stone raises as men crosses the age of forty years. Men are affected more than women. Diet and water mainly play role in the development of this ailment. Some species are found in use to treat this painful disorder. But in severe cases, people opt for treatment from professionals. If they are in the earlier phase of disorder, the local people try to treat with local remedies. The plant species in use should be examined for their virtues. Their scientific examinations will certainly help improvise their efficacy and safety.

Agriculture being the principal occupation, a large number of domestic animals are reared. They are also victims of different diseases or they met with accidental injuries. There is, no doubt, a precarious condition of medical facilities in this area for animal healthcare. Hence, again the local populace has to rely upon the locally available plant sources. They are employed in the form of different recipes, which are noted earlier also for human-beings. This study revealed about 44 species of angiospermic plants is used for combating their ailments. Nearly all types of plant parts, their products like latex or entire plants are beneficial.

Majority of population of this region live in rural parts or villages, where medical facilities are unavailable, either due to non-availability of expert medical persons or cost effectiveness of medicine. So these people solve their health problems in traditional ways. Although some live in urban area, there are also populations in slum areas which fight for bread and butter. They also can not afford the so called modern medical facilities. These usually find some other ways of treatment. Senior men and women are of great help in such circumstances. They mostly use substances available in the kitchen and in their surroundings. They are also aware of avoidances of drug administrations during certain periods like menstrual cycle, pregnancy, post-delivery period, menopause etc. in case of ladies. The age factor is also emphasized while prescribing some medicine. Accordingly they lower down the quantity and number of doses e.g. children and aged person are treated differently from other age groups. Consumption of drugs after dinner is mostly advised. The habits of drinking beverages like tea, coffee, wine and smoking are taken into consideration.

The beliefs and faith of these medicine–men, women and their society in general seem to be deeply associated with their medicaments. A literature survey on ethnomedicine even in other parts of our country indicates similar beliefs and faith. Their medicines are based on different causes of illness e.g. (a) angry deities who punish the wrongdoers (b) ghosts and ancestors feeling belittled, (c) sorcereors and witches hired by revenge seekers (d) loss of soul (e) possession by spirit (f) evil eye etc.

During the present study, it was noted that the inhabitants especially of farming communities in rural parts consume certain wild plants in raw or cooked form. Flowers and fruits are usually consumed raw, while the leaves and other vegetative parts are mostly cooked. These particularly form principal ingredients in their diets usually during famine periods and during normal seasons these are used as complementary to main food-stuff. Few of these are weeds gathered from cultivated fields. They use as many as 49 species as edibles (Appendix: IX).

It is further observed that as many as 169 wild species and crop weeds are used as fodder for domestic animals (Appendix: VI). Of these, few taxa are arborescent, whereas others are herbs, shrubs or climbers.

Some are made available to the farmers in their agricultural operations, while some are purposefully gathered by them. The people also have comprehensive knowledge about fodder-yielding species. Sometimes, farmer experience hardship in getting the fodder in the area. The area is monsoon-fed and hence usually during the rainy season, the resources are rich as compared to other seasons. Also, there is less availability of irrigated crop lands in this region. During such periods, perennial species are useful. To meet the demands, development of fodder banks, raising of grasslands with suitable palatable species and that too at the affordable costs are some of the necessities of the local people. It should be also interesting to know whether the species favoured for animal fodder have a specific nutritional significance. These should be chemically analyzed.

Investigators also revealed about 40 species used for fuel-wood (Appendix: V). of these eleven species are under cultivation for other purposes. The demand of fuel-wood is always increasing in rural areas. Wood still continues to be the main energy source. The situation is aggravated by the forces of denudation of forests areas everywhere. To redeem this situation, it always appears better to have captive plantations especially for fuel-wood. Patil (1999) recommended about 42 promising tree species suitable for energy plantations in the adjacent areas of Khandesh regions of Maharashtra. These can also be adapted for Buldhana district as suitable species because both regions fall under the nearly similar environmental conditions. It should be kept in mind that tree species with multiple uses should be planted covering indigenous as well as proper exotic species.

The houses are generally of 'dhaba' type with flat roof. Teak is the principal timber species employed for their construction. Some raise huts as their abodes. Apart from teak, they employ other timber yielding species (Appendix: VII). Few species are used especially for thatching roofs or partition walls. About 57 species are found commonly in vogue in this region. Few of them are under cultivation for some other purposes. Teak is also principal timber for making agricultural implements and tools or some form of furniture and. fixtures. Small timber is also obtained from some species for these purposes (Appendix: VIII).

The local inhabitants of different religions and castes use plants or their parts for various ceremonies, worships, festivals, births and deaths, funeral rites etc. These are mostly derived from cultivated plants. Wild species are also noted by the present investigation e.g.

Aegle marmelos (it is planted also), *Salvadora persica, Dinebra retroflexa* and few others.

A large number of species are employed for the purpose of human medicine.This indicates that the area is a reservoir of ethnomedicinally useful plants. Alcorn (1981) opined that once a species gets known as a resource in any human society, the impact of this knowledge on expansion, distribution, threat to that species and in cases even its extinction play a great role. Thus conservation and intervention of science and technology is desired. The subsequent scientific examinations should be geared up in a large scale throughout.

The ethnomedicinal plantlore is regaining its lost reputation. The green wave all over the country and the world is putting long strides. This will lead remarkable discoveries from the word of plant-based ethnomedicines. The appreciation and efforts of the phytochemists, pharmacologists along with the studies of biological activities and clinical trials would help survive the golden heritage of our wise ancestors. Their researches will decisively bring new leads or molecules to a forefront for the well-being of the human society at large.

Traditionally, some wild plant species find place in the diets of rural people. Such species are in addition to the cultivated ones. They also play a great role in the economy of rural. They are cherished particularly during famine periods or occasionally otherwise. Some species produce leafy vegetables and fruits. About eight species are reported for this purpose (Appendix: IX).

In these different tehsils of district, about 412 species play a role in the welfare of mankind in respect of food, medicine and shelter, apart from their agricultural produce. Plants are real benefactors. The inhabitants of the area under investigation struggle to have healthy living. They develop in-depth knowledge and are traditionally passed to the next generations with certain refinements or even omissions. Their traditional treasures, with certain value-additions are exploited, nay sometimes over-exploited by the civilized world.

In modern times, the forces of industrialization, urbanization and the developmental activities are held responsible for depleting the ambient biota. At the same time, the process of acculturation of the ethnic societies is depriving away the traditional treasure of plantlore at an alarming rate. It is now high time that all the government and non-government programmes should intensively double their efforts to rescue from these devastating forces.

It is, therefore, suggested in the light of the present study that medicinal plant conservation or farming should taken up in all open forests areas or barren lands. Timber and fuel-food is another necessity of the rural people. To meet their needs, captive plantations on rotational basis should be favoured with people's participation. The supply of fodder is also not adequate and regular for rearing pet animals. Fodder banks should be established for its timely supply to the needy. Grazing lands should be maintained. These basic needs of the rural people are to be satisfied, apart from the food.

The traditional plantlore divulged during the course of investigation from this area will be useful to the phytochemists, druggists, industrialists, foresters, ecologists for management of developments, apart from teachers, students and research workers in botany, ethnobotany, anthropology and medicine for their teaching and research.

Summary

The traditional plantlore in various tehsils of Buldhana district of Vidarbha region (Maharashtra) is investigated. Attempts have been made to garner information on intimate relationships between man and plant species in all corners of the area under investigation. The data has been tapped particularly on ethnomedicines, both for human beings and domestic animals, apart from uses on food, fodder, fibre, oils, timber, agricultural implements and tools etc. Few plants of religious significance are also noted. During the frequent visits paid, pertinent information on their culture, life style, beliefs, etc. have been documented. The administrations of plant drugs or their products as medicine for mankind and the domestic animals are critically recorded.

The first-hand information is collected for 412 species belonging 91 families of angiosperms and pteridophytes as principal drugs, apart from other supplementary 74 species. Of these, 58 species are crop weeds. The fairly rich plant wealth provides inhabitants enormous unconventional resources for different necessities of human life. To date, information on the uses and applications of plants for varied purposes was passed from mouth to mouth. The investigator has made pioneering work in this part of state. There are many species which are used multipurposely.

Majority of species are in vogue for medicinal purpose of mankind, whereas as many as 90 species are found commonly employed in the medicaments of the domestic animals. The other uses are recorded for fodder (169 species), fuel-wood (40 species), house and hut constructions (57 species), agricultural implements (16 species), antidotes (31 species) and some others for miscellaneous purposes.

The disease or ailments related to head, ears, eyes, nose, throat, respiratory tract, digestive and nervous systems, skin, bone fracture, urinary and genital organs, bites or stings of snake, scorpion, dog, common fever, typhoid, malaria, ulcers of mouth, general body ache, diabetes, etc. are cured by folk medicines available locally. They mostly employ recipes like decoction, infusion, extract, juice, paste, powder, fumes, oil, pellets etc. Apart from the plants or their parts, they add certain substances like milk, curd, whey, oils, sugar, honey, ant-hill soil, animal ghee, cow-urine, goat-urine, jaggery and few others available in their kitchens or surroundings.

The investigators thought it desirable to extend scientific and technological studies like chemical screening, pharmacological investigation, biological activities and clinical trials on modern lines. The results of such studies will help resolve the problems of population explosion, inadequate supply and alternative sources of drugs, high cost of treatments, side effects of drugs etc. Collaborative research involving experts from different disciplines will yield better results. Moreover, documentation of traditional medicines and other form of uses will help us for patenting in future. The investigator also suggests measures for plant conservation, fodder bank, and captive plantations especially for fuel-wood.

This work also sheds light on geography of the district, area (tehsils) under investigation, geology, soils, forests, climate and seasons, history, the people of area, plant based industries, crops in the region, botanisation in the region etc. to back up fully the topic of investigation. The investigators feel that this contribution will be useful for those engaged in the researches in food, medicine and other forms of useful products.

APPENDICES

APPENDIX-I
Plants Used For Veterinary Medicine

Sl.No.	Plant Name	Sl.No.	Plant Name
1.	*Abelmoschus manihot*	2.	*Acalypha indica*
3.	*Acanthospermum hispidums*	4.	*Achyranthes aspera*
5.	*Adansonia digitata*	6.	*Aegle marmelos*
7.	*Aerva lanata*	8.	*Ailanthus excela*
9.	*Amaranthus blitum*	10.	*Amaranthus spinosus*
11.	*Ampleocissus latifolia*	12.	*Annona squamosa*
13.	*Aristolochia bracteolata*	14.	*Asparagus racemosa*
15.	*Balanites aegyptiaca*	16.	*Basella alba*
17.	*Biophytum sensitivum*	18.	*Bombax ceiba*
19.	*Boswellia serrata*	20.	*Butea monosperma*
21.	*Cajanus cajan*	22.	*Calatropis gigantea*
23.	*Calatropis procera*	24.	*Capparis zeylanica*
25.	*Cardiospermum halicacabum*	26.	*Careya arborea*
27.	*Cassia obtusifolia*	28.	*Cassia occidentalis*
29.	*Cassine ablens*	30.	*Celosia argentia*
31.	*Cissus quandrangula*	32.	*Citrus aurantifolia*
33.	*Citrus sinensis*	34.	*Cleome gynandra*
35.	*Clitoria biflora*	36.	*Coccinia indica*
37.	*Commelina forskalaei*	38.	*Convolvulus arvensis*
39.	*Cordia gharaf*	40.	*Coriandum sativum*
41.	*Cryptolepis buchanani*	42.	*Cucumis calossus*
43.	*Cucurbita moschata*	44.	*Cucurbita pepo*
45.	*Cullen corylfolia*	46.	*Curcuma longa*
47.	*Cuscuta chinensis*	48.	*Cymbopogan martini*
49.	*Datura innoxia*	50.	*Eclipta prostrate*
51.	*Ehretia aspera*	52.	*Euphorbia indica*
53.	*Euphorbia tirucalli*	54.	*Ficus bengalensis*
55.	*Helicteres isora*	56.	*Holarrhena pubescens*
57.	*Hordeum vulgare*	58.	*Ipomoea fistulosa*

Contd...

APPENDIX-I–Contd...

Sl.No.	Plant Name	Sl.No.	Plant Name
59.	*Ipomoea hederifolia*	60.	*Kedrostis rostrata*
61.	*Lagerstroemeria parviflora*	62.	*Launea procumbens*
63.	*Leucaena latisiliqua*	64.	*Lindernia parviflora*
65.	*Manikara zapota*	66.	*Momordica dioica*
67.	*Morinda pubescens*	68.	*Musa paradisiaca*
69.	*Nerium indicum*	70.	*Nicotiana tabacum*
71.	*Ocimum basilicum*	72.	*Ocimum tenuiflorum*
73.	*Pergularia daemia*	74.	*Physalis minima*
75.	*Punica granatum*	76.	*Pupalia lappacea*
77.	*Ricinus communis*	78.	*Salvadora persica*
79.	*Semecarpus anacardium*	80.	*Solanum virginianum*
81.	*Tabernaemontana varticata*	82.	*Tamarindus indica*
83.	*Tectona grandis*	84.	*Terminalia arjuna*
85.	*Terminalia bellirica*	86.	*Thespesia populnea*
87.	*Tribulus lanuginous*	88.	*Ventilago denticulata*
89.	*Vernonia cinerea*	90.	*Withnia somnifera*

APPENDIX-II
Supplementary Medicinal Plants

Sl.No.	Plant Name	Sl.No.	Plant Name
1.	*Azadirachta indica*	2.	*Acacia catechu*
3.	*Acacia chundra*	4.	*Acacia nilotica ssp. indica*
5.	*Allium sativum*	6.	*Arachis hypogaea*
7.	*Asparagus racemosus*	8.	*Azadirachta indica*
9.	*Balanites aegyptiaca*	10.	*Caesalpinia bonduc*
11.	*Cajanus cajan*	12.	*Calotropis procera*
13.	*Canthium coromandelicum*	14.	*Capsicum annuum*
15.	*Catharanthus roseus*	16.	*Cicer arientinum*
17.	*Cissampelos pareira*	18.	*Citrus aurantifolia*
19.	*Cocos nucifera*	20.	*Corchorus depressus*
21.	*Cucumis calossus*	22.	*Cuminum cyminum*
23.	*Curcuma amada*	24.	*Curcuma longa*
25.	*Cuscuta chinensis*	26.	*Dolichandrone falcata*
27.	*Ellettaria cardamomum*	28.	*Embelia tsjeriam-cottam DC.*
29.	*Emblica officinalis*	30.	*Eucalyptus maculate*
31.	*Ficus amplissima*	32.	*Ficus bengalensis*
33.	*Glycyrrhiza glabra*	34.	*Helianthus annuus*
35.	*Ipomoea carnea*	36.	*Justicia adhthoda*
37.	*Lawsonia inermis*	38.	*Madhuca longifolia*
39.	*Mangifera indica*	40.	*Mentha piperita*
41.	*Momordica charantia*	42.	*Mucuna pruriens*
43.	*Nardostachys jatamansi*	44.	*Nicotiana tabacum*
45.	*Ocimum kilimandscharicum*	46.	*Ocimum tenuiflorum*
47.	*Oryza sativa*	48.	*Pennisetum americanum*
49.	*Phoenix dactylifera*	50.	*Piper betle*
51.	*Piper nigrum*	52.	*Pongamia pinnata*
53.	*Ricinus communis*	54.	*Saccharum officinarum*
55.	*Santalum album*	56.	*Sapindus emarginata*
57.	*Sesamum orientale*	58.	*Sorghum bicolor*

Contd...

APPENDIX-II–Contd...

Sl.No.	Plant Name	Sl.No.	Plant Name
59.	*Spilanthes calva*	60.	*Syzigium cumini*
61.	*Syzygium aromaticum*	62.	*Tamarindus indica*
63.	*Tectona grandis*	64.	*Terminalia bellirica*
65.	*Tinospora cordifolia*	66.	*Trachyspermum ammni*
67.	*Trigonella foenum-graecum*	68.	*Triticum aestivum*
69.	*Vernonia anthelmintica*	70.	*Vitex negundo*
71.	*Vitis vinifera*	72.	*Withania somnifera*
73.	*Zingiber officinale*	74.	*Ziziphus mauritiana*

APPENDIX-III
Plants Used For Antidotes

Sl.No.	Plant Name	Sl.No.	Plant Name
1.	*Acacia leucophloea*	2.	*Achyranthes aspera*
3.	*Allium cepa*	4.	*Amorphophallus commutatus*
5.	*Apluda mutica*	6.	*Argemone mexicana*
7.	*Aristolochia bracteolata*	8.	*Azadirachta indica*
9.	*Balanites aegyptiaca*	10.	*Calotropis procera*
11.	*Canthium coromandelicum*	12.	*Capsicum annuum*
13.	*Cassia obtusifolia*	14.	*Cassia tora*
15.	*Cassine albens*	16.	*Citrullus lanatus*
17.	*Cucumis callossus*	18.	*Cymbopogan martini*
19.	*Cyperus rotundus*	20.	*Datura innoxia*
21.	*Elettaria cardamomum*	22.	*Ficus religiosa*
23.	*Gossypium hirsutum*	24.	*Kedrostis rostrata*
25.	*Luffa acutangula*	26.	*Mangifera indica*
27.	*Plumeria alba*	28.	*Sapindus emarainatus*
29.	*Solanum virginianum*	30.	*Tribulus lanuginosus*
31.	*Vitis vinifera*		

APPENDIX-IV
Crop Weeds

Sl.No.	Plant Name	Sl.No.	Plant Name
1.	*Acalypha malabarica*	2.	*Achyranthes aspera*
3.	*Aerva lananta*	4.	*Ageratum conyzoides*
5.	*Alternanthera tenella*	6.	*Alysicarpus longifolius*
7.	*Amaranthus polygonoides*	8.	*Amaranthus spinosus*
9.	*Argemone mexicana*	10.	*Aristolochia bracteolata*
11.	*Asphodelus tenuifolius*	12.	*Biophytum sensitivum*
13.	*Cassia obtusifolia*	14.	*Cassia tora*
15.	*Celosia argentia*	16.	*Chenopodium murale*
17.	*Chloris virgata*	18.	*Cleome viscosa*
19.	*Cocculus hirsutus*	20.	*Commelina bengalensis*
21.	*Commelina forskalaei*	22.	*Convolvulus arvensis*
23.	*Corchorus depressus*	24.	*Corchorus olitorius*
25.	*Corchorus trilocularis*	26.	*Cucumis calossus*
27.	*Cyanotis cristata*	28.	*Cynadon dactylum*
29.	*Cyperus rotundus*	30.	*Dactyloctinum aegyptium*
31.	*Datura innoxia*	32.	*Datura metel*
33.	*Desmodium dichotomum*	34.	*Desmodium velutinum*
35.	*Digera muricata*	36.	*Dinebra retroflexa*
37.	*Echinops echinatus*	38.	*Euphorbia indica*
39.	*Euphorbia prunifolia*	40.	*Launaea procumbens*
41.	*Leucas cephalotes*	42.	*Ocimum americanum*
43.	*Phyllanthus airy-shawii*	44.	*Phyllanthus amarus*
45.	*Phyllanthus maderaspatensis*	46.	*Physalis minima*
47.	*Portulaca oleracea*	48.	*Setaria verticillata*
49.	*Solanum nigrum*	50.	*Solanum virginianum*
51.	*Sonchus oleraceus*	52.	*Striga asiatica*
53.	*Sporobolus coromadelianus*	54.	*Tridax procumbens*
55.	*Triumfetta rhomboidea*	56.	*Vernonia cinerea*
57.	*Xanthium indicum*		

APPENDIX-V
Plant Used As Fuel

Sl.No.	Plant Name	Sl.No.	Plant Name
1.	*Abelmoschus manihot*	2.	*Abutilon indicum*
3.	*Acacia chundra*	4.	*Acacia leucophloea*
5.	*Acacia nilotica*	6.	*Ailanthus excelsa*
7.	*Annona squamosa*	8.	*Anogeissus latifolia*
9.	*Bombax ceiba*	10.	*Butea monosperma*
11.	*Cajanus cajan*	12.	*Citrus sinensis*
13.	*Cleodendron multiflorum*	14.	*Cordia gharaf*
15.	*Dalbergia sissoo*	16.	*Delonix elata*
17.	*Delonix regia*	18.	*Gossypium hirsutum*
19.	*Hardwickia binata*	20.	*Holarrhena pubescens*
21.	*Lagerstroemeria parviflora*	22.	*Lantana camara*
23.	*Leucaena latisiliqua*	24.	*Limonia acidissima*
25.	*Mangifera indica*	26.	*Manikara zapota*
27.	*Melia azedarch*	28.	*Moringa concanensis*
29.	*Pithecellobium dulce*	30.	*Prosopis cineraria*
31.	*Psidium guajava*	32.	*Ricinus communis*
33.	*Semecarpus anacardium*	34.	*Sesamum orientale*
35.	*Sesbania grandiflora*	36.	*Sesbania sesban*
37.	*Tamarindus indica*	38.	*Terminalia bellirica*
39.	*Terminalia catappa*	40.	*Xanthium indicum*

APPENDIX-VI
Plants Used As Fodder

Sl.No.	Plant Name	Sl.No.	Plant Name
1.	*Abelmoschus manihot*	2.	*Abutilon indicum*
3.	*Abutilon indicum*	4.	*Acacia leucophloea*
5.	*Acacia nilotica*	6.	*Acaypha malabarica*
7.	*Achyranthus aspera*	8.	*Adansonia digitata*
9.	*Aerva lanata*	10.	*Ageratum conyzoides*
11.	*Albizia lebbeck*	12.	*Alternanthera sessilis*
13.	*Alternanthera tenella*	14.	*Alysicarpus longifolius*
15.	*Alysicarpus tetragonolobus*	16.	*Amaranthus viridis*
17.	*Amaranthus polygonoides*	18.	*Anogeissus latifolia*
19.	*Apluda mutica*	20.	*Azadirachta indica*
21.	*Balanites aegytiaca*	22.	*Bauhinia racemosa*
23.	*Biophytum sensitivum*	24.	*Boerhavia diffusa*
25.	*Boerhavia erecta*	26.	*Boerhavia repens*
27.	*Butea monosperma*	28.	*Caesalpinia decapetala*
29.	*Cajanus cajan*	30.	*Capparis grandis*
31.	*Capparis sepiaria*	32.	*Capparis zeylanica*
33.	*Cardiospermum halicacabum*	34.	*Carrisa congesta*
35.	*Cassia arborea*	36.	*Cassia fistula*
37.	*Cayratia trifolia*	38.	*Celosia argentia*
39.	*Chenopodium murale*	40.	*Chloris virgata*
41.	*Cissampelos pareira*	42.	*Cleodendron multiflorum*
43.	*Clitoria ternatea*	44.	*Coccinia grandis*
45.	*Coccinia indica*	46.	*Cocculus hirsutus*
47.	*Coix gigantia*	48.	*Combretum albidum*
49.	*Commelina bengalensis*	50.	*Commelina forskalaei*
51.	*Commicarpus chinensis*	52.	*Convolvulus arvensis*
53.	*Corchorus olitorius*	54.	*Corchorus trilocularia*
55.	*Cordia dichotoma*	56.	*Cordia gharaf*
57.	*Crotalaria filipes*	58.	*Cryptolepsis buchanani*

Contd..

APPENDIX-VI–Contd...

Sl.No.	Plant Name	Sl.No.	Plant Name
59.	*Cryptostegia grandiflora*	60.	*Ctenolepis garcinii*
61.	*Cucumis callosus*	62.	*Cucumis melo*
63.	*Cucurbita maxima*	64.	*Cullen corylifolia*
65.	*Cyanotis cristata*	66.	*Cynadon dactylum*
67.	*Cyperus bulbosus*	68.	*Cyperus nutans*
69.	*Dactyloctenium aegptium*	70.	*Dalbergia latifolia*
71.	*Delonix elata*	72.	*Desmodium dichotomum*
73.	*Desmodium gangeticum*	74.	*Desmodium velutinum*
75.	*Dichanthium annulatum*	76.	*Dichanthium foveolatum*
77.	*Digera muricata*	78.	*Dinebra retroflexa*
79.	*Dinebra retroflexa*	80.	*Dioscorea bulbifera*
81.	*Diplocyclos palmatus*	82.	*Dolichandrone falcata*
83.	*Elephantopus scaber*	84.	*Elyonurus royleanus*
85.	*Emblica officinalis*	86.	*Emilia sonchifolia*
87.	*Euphorbia indica*	88.	*Euphorbia prunifolia*
89.	*Ficus nervosa*	90.	*Flacourtia indica*
91.	*Gmelina arborea*	92.	*Goniogyna hirta*
93.	*Grewia flavescens*	94.	*Grewia tiliaefolia*
95.	*Hardwickia binata*	96.	*Hemidesmus indicus*
97.	*Heteropogan contortus*	98.	*Holoptelea integrifolia*
99.	*Hordeum vulgare*	100.	*Ipomoea hederifolia*
101.	*Ipomoea nil*	102.	*Ipomoea obscura*
103.	*Kirganelia reticulata*	104.	*Langenaria siceraria*
105.	*Launaea procumbens*	106.	*Leucaena latisiliqua*
107.	*Leucas cephalotes*	108.	*Limonia acidissima*
109.	*Maerua oblongifolia*	110.	*Mangifera indica*
111.	*Maytenus senegalensis*	112.	*Merremia aegyptia*
113.	*Mimosa hamata*	114.	*Mitragyna parvifolia*
115.	*Moringa oleifera*	116.	*Mukia maderaspatana*
117.	*Musa paradisiaca*	118.	*Ougeinia oojenensis*
119.	*Oxalis corniculata*	120.	*Pennisetum purpureum*

Contd...

APPENDIX-VI–Contd...

Sl.No.	Plant Name	Sl.No.	Plant Name
121.	*Pentanema indicum*	122.	*Phyllanthus air-shawi*
123.	*Phyllanthus airy-shawii*	124.	*Phyllanthus amarus*
125.	*Phyllanthus maderaspatensis*	126.	*Pithecellobium dulce*
127.	*Polygonum glabrum*	128.	*Portulaca oleracea*
129.	*Portulaca quadrifida*	130.	*Prosopis cineraria*
131.	*Pupalia lapacea*	132.	*Rhus sinuata*
133.	*Rhus sinuata*	134.	*Rhynchosia minima*
135.	*Rivea hypocrateriformis*	136.	*Rothboalia exaltata*
137.	*Rottboellia cochinchinensis*	138.	*Saccharum officinarum*
139.	*Salvadora persica*	140.	*Santalum album*
141.	*Schleichera oleosa*	142.	*Sclerocarpus Africanus*
143.	*Securinega leucopyrus*	144.	*Securingea virosa*
145.	*Sesbania bispinosa*	146.	*Sesbania grandiflora*
147.	*Sesbania seban*	148.	*Setaria verticillata*
149.	*Sida acuta*	150.	*Sida spinosa*
151.	*Solanum nigrum*	152.	*Sonchus oleraceus*
153.	*Sporobolus coromandelianus*	154.	*Sporobolous*
155.	*Stylosanthes fructiosa*	156.	*Syzigium cumini*
157.	*Tamarindus indica*	158.	*Terminalia bellirica*
159.	*Terminalia catappa*	160.	*Terminalia chebula*
161.	*Themeda quadrivalvis*	162.	*Tinospora cordifolia*
163.	*Trema orientalis*	164.	*Trichodesma zeylanicum*
165.	*Triumfetta annua*	166.	*Wattakaka volubilis*
167.	*Ziziphus mauritiana*	168.	*Ziziphus nummularia*
169.	*Zizipus glabrata*		

APPENDIX-VII
Plants Employed For Huts And Houses

Sl.No.	Plant Name	Sl.No.	Plant Name
1.	*Abelmoschus manihot*	2.	*Abutilon indicum*
3.	*Acacia chundra*	4.	*Acacia nilotica ssp. indica*
5.	*Ailanthus excelsa*	6.	*Apluda mutica*
7.	*Azadirachta indic*	8.	*Bambusa arundinacea*
9.	*Buchanania lanzan*	10.	*Butea monosperma*
11.	*Cajanus cajan*	12.	*Capparis zeylanica*
13.	*Cassia obtusifolia*	14.	*Cassia tora*
15.	*Cleodendron multiflorum*	16.	*Cleome viscosa*
17.	*Cocos nucifera*	18.	*Combretum albidum*
19.	*Cordia gharaf*	20.	*Cullen corylifolia*
21.	*Dalbergia latifolia*	22.	*Dalbergia sissoo*
23.	*Delonix elata**	24.	*Dinebra retroflexa*
25.	*Eucalyptus camaldulensis*	26.	*Eucalyptus maculata**
27.	*Flacourtia indica*	28.	*Gliricidia sepium**
29.	*Gmelina arborea*	30.	*Gossypium hirsutum**
31.	*Hardwickia binata*	32.	*Helianthus annuus**
33.	*Hibiscus caesius*	34.	*Hibiscus rosa-sinensis*
35.	*Ipomoea carnea*	36.	*Lantana camara*
37.	*Leucaena latisiliqua*	38.	*Limonia acidissima**
39.	*Mangifera indica**	40.	*Melia azedarach*
41.	*Mitragyna parvifolia*	42.	*Moringa concanensis*
43.	*Prosopis cineraria*	44.	*Saccharum officinarum**
45.	*Saraca asoca**	46.	*Semecarpus anacardium*
47.	*Sesamum orientale*	48.	*Sesbania bipinosa*
49.	*Setaria verticillata*	50.	*Tamarindus indica*
51.	*Tectona grandis*	52.	*Terminalia arjuna*
53.	*Terminalia bellirica*	54.	*Terminalia catappa**
55.	*Terminalia crenulata*	56.	*Thespesia lampas*
57.	*Vitex negundo*		

APPENDIX-VIII
Plants Used For Agricultural Implements

Sl.No.	*Plant Name*	*Sl.No.*	*Plant Name*
1.	*Acacia leucophloea*	2.	*Acacia nilotica*
3.	*Azadirachta indica*	4.	*Butea monosperma*
5.	*Cordia gharaf*	6.	*Cordia macleodii*
7.	*Dalbergia sissoo*	8.	*Eucalyptus maculata*
9.	*Hardwickia binata*	10.	*Mangifera indica*
11.	*Ougeinia oojenensis*	12.	*Tectona grandis*
13.	*Terminalia arjuna*	14.	*Terminalia bellirica*
15.	*Terminalia crenulata*	16.	*Zizipus glabrat*

APPENDIX-IX
Plants Used As Fruits And Vegetables

Sl.No.	Plant Name	Sl.No.	Plant Name
1.	*Abelmoschus manihot*	2.	*Amaranthus spinosus*
3.	*Amaranthus polygonoides*	4.	*Amaranthus viridus*
5.	*Anoegeissus latifolia*	6.	*Balanites aegyptiaca*
7.	*Bambusa aurandinacea*	8.	*Basella alba*
9.	*Butea monosperma*	10.	*Capsicum annuum*
11.	*Carissa congesta*	12.	*Carthamus tinctorius*
13.	*Cassia tora*	14.	*Clitoria biflora*
15.	*Colocasia esculenta*	16.	*Commelina bengalensis*
17.	*Commelina forskalaei*	18.	*Corchorus olitorius*
19.	*Cordia dichotoma*	20.	*Cordia gharaf*
21.	*Cordia macleodii*	22.	*Cucumis melo*
23.	*Cucurbita moschata*	24.	*Digera muricata*
25.	*Emblica officinalis*	26.	*Ficus nervosa*
27.	*Helicteres isora*	28.	*Hibiscus caesius*
29.	*Holoptelea integrifolia*	30.	*Kigella africana*
31.	*Kirganelia reticulata*	32.	*Lagenaria siceraria*
33.	*Limonia acidissima*	34.	*Momordica charantia*
35.	*Mukia maderaspatana*	36.	*Murraya oleifera*
37.	*Musa paradisiaca*	38.	*Opuntia elatior*
39.	*Oxalis corniculata*	40.	*Pithecellobium dulce*
41.	*Pongamia pinnata*	42.	*Portulaca oleracea*
43.	*Portulaca quadrifida*	44.	*Rhus sinuata*
45.	*Rivea hypocrateriformis*	46.	*Solanum nigrum*
47.	*Sphaeranthus senegalensis*	48.	*Tamarindus indica*
49.	*Zizipus glabrata*		

APPENDIX-X
Botanicals Purchased

Sl.No.	Plant Name	Part Purchased
1.	*Acacia catechu*	Catechu
2.	*Brassica juncea*	Seeds
3.	*Chlorophytum borivilianum*	Root Tubers
4.	*Cocos nucifera*	Fruits
5.	*Coffea Arabica*	Seeds
6.	*Coriandrum sativum*	Seeds
7.	*Cuminum cyminum*	Seeds
8.	*Curcuma amada*	Rhizomes
9.	*Curcuma aromatica*	Rhizome
10.	*Curcuma longa*	Rhizomes
11.	*Cymbopogon martinii*	Leaf oil
12.	*Elettaria cordamomum*	Seeds
13.	*Gompherna globosa*	Fruits
14.	*Holarrhena pubescens*	Seeds
15.	*Nardostachys jatamansi*	Roots
16.	*Nicotiana tabacum*	Leaf pieces
17.	*Ocimum kilimandscharium*	Camphor
18.	*Papaver somniferum*	Seeds
19.	*Phoenix dactylifera*	Fruits
20.	*Piper betle*	Leaves
21.	*Piper nigrum*	Seeds
22.	*Sapindus emarginatus*	Fruits
23.	*Selaginella ciliaris*	Whole plants
24.	*Semecarpus anacardium*	Seeds
25.	*Spilanthes calva*	Flowers
26.	*Syzgygium aromaticum*	Dried flower buds
27.	*Terminalia chebula*	Fruits
28.	*Trachyspermum ammi*	Seeds
29.	*Trigonella foenum-graceum*	Seeds
30.	*Vitis vinifera*	Fruits
31.	*Zingiber officinale*	Rhizomes

References

Alcorn, J.B. (1981) Factors influencing botanical resource perception among the Haustec: Suggestion for future ethnobotanical inquiry. J. Ethnobiol . 1: 221-220.

Anonymous (1976) Gazetteer of India, Maharashtra State Gazetteers, Buldhana District. Gazetteers Department, Government of Maharashtra, Bombay, India.

Badhe, P.D. and B.N. Sharma (1981) Ethnomedicinal methods of Korku tribes of west Melghat. B. M. E. B. R. 3(2-4): 162-172.

Badhe, P.D. and V. K. Pandey (1988) A study of medicinal and economic plants of Amaravati Division, Amaravati Circle, Maharashtra B. M. E. B. R. 11(1-4): 1-39.

Badhe, P.D., B.N. Sharma and V. K. Pandey (1997)Rare folk medicines practiced by Gond tribes of Wadsa forest division, Gadchiroli Dist t (Maharashtra State). Bull. Medico-ethno-Bot: Res. 18(1-2): 12-17.

Balapure, K. M. (1966) Some plant records from the erstwhile C.P.and Berar. Journ. Bombay Nat. Hist. Soc. 62: 455-462.

Bank, T. P. – II (1957 a) Report on the Sub-Committee on Ethnobotany. Proc.9th Pacific Sci. Cong. (Bangkok) 4: 21-22.

Bank, T. P. –II (1957 b) Openning remarks in ethnobotany and ecology. Proc. 9th Pacific Sci. Cong. (Bangkok) 4: 244—245.

Bhogaonkar, P. Y. and V.D. Devakar (2001 a) Studies in ethnobotany of Korkus of Melghat Amaravati District, Maharashtra, I-Antisterility Drugs. JAST 4:(1-2):63-66.

Bhogaonkar, P. Y. and V. D. Devarkar (2001 b) Studies in ethnobotany of Korkus of Melghat, Amaravati District, Maharashtra. II. Drugs for urino-genital disorders. JAST 4 (1-2) 67-68.

Bhogaonkar, P.Y. and V.N. Kadam (2006) Etho-medicobotany of some local reproductive health practices in Umakhed of Yavatmal district in Maharashtra. Ethnobotany 18: 107-113.

Bhogaonkar, P. Y. and V. D. Devarkar (2002) Some unique etnomedicinal plants of Korkus of Melghat Tiger Reserve Maharashtra, Ethnobotany 14: 16-19.

Bodding, P. O. (1925,1927) Studies on Santal medicines and enacted folklore. Part I and II. Mem. Asiat. Soc. Bengal 10: 1-427.

Brandis, D. (1974) The forest flora of North-West Central India, London, U.K.

Bye, R.A. (1985) Botanical perspectives of ethnobotany of the Greater Siuthwest. Econ. Bot. 39:375-386.

Chaudhari, U. S. and Varsha Hutke (2002) Ethno-medico botanical information on some plants used by Melghat tribes of Amaravati district, Maharashtra, Ethnobotany 14: 100-102.

Chute, G.S. and V.J. Tiwari (1999) Indigenous ethnomedicinal plants used by tribal people of Buldhana and Gadchiroli districts of Maharashtra state, Indian J. Nat. Prod. 15 (1): 3-8.

Cox, P. A. (2005) The seven pillars of ethnobotanical wisdom. Ethnobotany 17: 24-34.

Dande, V. P. (1966) *Scirpus kerni* Raymond- a new record for India. Bull. Bot, Surv. India. 8: 358.

Das, P. K. and M.K. Mishra (1988) Some medicinal plants among Kondhas around Chandrapur (Koraput) J. Econ. Tax. Bot. 12 (1): 103-109.

Deshmukh, V.R. and M.A. Dhore (1994) Nutritive values of some lesser known herbivor food from Melghat tiger reserve Maharashtra. The Indian forester 120 (10): 920-923.

Deshpande, B. D. and M. B. Deshpande (1963) Flora of Chandmari hillocks, Wardha (Maharashtra) J. Indian Bot. Soc.42: 143-154.

Dhore, M.A. (2002) Flora of Amravati District with Special Reference to the Distribution of Tree Species, Registrar, Amaravati University, Amravati, India.

Diwakar, P.G. and B.D.Sharma (2000). Flora of Buldhana District Bot. Surv. India, Kolkata, India.

*Faulks, P. J. (1958) An Introduction to Ethnobotany. Moredale Publications, London, U.K.

Giri, R. B. (1994) Melghattil Gushadopyogi Vanaspati Technical Bulletin No. 5 Directorate, Project Tiger, Melghat Paratwada, District Amaravati, Maharashtra.

*Graham, R. J. D. (1911) List of wild plants found on the Nagpur and Tilikhedi farms. Nagpur.

Harshberger, J. W. (1885) Some New Ideas. Philadelphia Evening Telegraph.

*Harshberger, J. W. (1896) The purpose of Ethnobotany. Bot. Gaz. 21 (3): 146-154.

Heiser, C.B. (1985) Ethnobotany of Naranjilla (*Solanum quitoense*) and its relatives. Bot.Gaz. 39:4-11.

Jain, S. K. (1986) Ethnobotanical studies on medicinal plants. Curr.Res.Med.Pl.In India,INSA 1-5.

Jain, S. K. (1987) A Manual of Ethnobotany, Scientific Publishers Jodhpur.

Jain, S. K. (1989) Methods and approaches in Ethnobotany Soc. Ethnobotanists. Lucknow. India.

Jain, S. K. and R. R. Rao (1977) A Handbook of Field and Herbarium Methods, Today and Tomorrow Publishers New Delhi, India.

Jones, V. H. (1941) The nature and scope of ethnobotany. Chronica Botanica 6 (10): 215-221.

Kamble, S. K. and S. G. Pradhan (1982) Ethnobotany of the Korkus in Maharashtra Bull. Bot. Surv. India 22 (1-4): 201-202.

Kamble, S. K. and S. G. Pradhan (1988) Flora of Akola District (Maharashtra). Bot. surv. of India, Calcutta, India.

Kshirsagar, S. R. and Patil D.A. (2008) Forest Flora of Jalgaon District (Maharashtra). Bishan Singh Mahendra Pal Singh, Dehradun, India.

Lakshinarasimhan P. and B. D. Sharma (1991) Flora of Nasik District. Bot. Surv. India, Calcutta, India.

Malhotra, S. K. and Moorthy (1973) Some useful and medicinal plants of Chandrapur district, Maharashtra state, Bull. Bot. Serv. India. 15 (1& 2): 13-1.

Malhotra, S. K. and S. Moorty. (1971) Material for the flora of Chandrapur district, Maharashtra state. Bull. Bot. Surv. India 13: 292-318.

Manilal, K.S. (1989) Linkages of ethnobotany with other sciences and disciplines. Ethnobotany I: 15-24.

*Martin, G. J. (1995) Ethnobotany. Chapman and Hall, London, U.K.

Mashelkar, R. A. (2002) In: Forward. Indian J. Trad. Knowledge 1(1): 1.

Mirashi, M. V. (1954) Studies in the hydrophytes of Nagpur. J. Indian Bot. Soc. 33: 299-308.

Mirashi, M. V. (1960) Some new plants records for Nagpur. 1: 23-30. Bull. Bot. Soc. Coll, Sci. Nagpur.

Mirashi, M. V. and S. A. Paradkar (1962) Studies on the Compositae of Nagpur I. Some new plant records. Bull. Bot. Soc. Coll. Sci. Nagpur. 3 (2): 59-67.

Mirashi, M. V. and Usha Nafade (1961) Studies in the Tubifloeae of Nagpur I- Some new plant records. Bul. Bot. Soc. Coll. Sci. Nagpur 2: 9-20.

Naik, V. N. (1998) Flora of Maharashtra Vol.1-2. Amrut Prakashan, Aurangabad (M.S.), India.

Patel, R. I. (1968) Forest Flora of Melghat. Bishen Singh Mahendra Pal Singh, Dehradun, India.

Patil, D.A. (1999) In quest of suitable tree species for energy plantation in Khandesh region (Maharashtra) JAST 2 (1): 9-15.

Patil, D.A. (2003) Flora of Dhule and Nandurbar Districts (Maharashtra). Bishen Singh Mahendra Pal Singh, Dehradun, India.

Peter Paul Hembrom (1996) Contact therpy practiced by Mundas of Chotanagpur (Bihar). Ethnobotany 8: 36-39.

Punjani, Bhaskar L. (2002) Plants used in contact therpy by tribals of Sabarkantha (Gujarat). Ethnobotany 14:57-59.

*Purekar, P. N. (1985) Grasses of Nagpur District-Taxonomy and Palynology. Ph.D. Thesis. Nagpur.

Richard, I. Ford (1978) The nature and status of ethnobotany. Ann. Arbor. Michigan.

*Schultes, R. E. (1941) Lethnobotanica on alcancoy Sus objects. Cadesia 3: 7-12.

Schultes, R. E. (1960) Tapping our heritage of ethnobotanical lore. Econ. Bot. 14: 257-263.

Schultes, R. E. (1962) The role of ethnobotanist in the search of new medicinal plants. Lloydia 25: 257-266.

Sharma, B. D., S. Karthikeyan and N. P. Singh (1996) Flora of Maharashtra State: Monocotyledons. BSI, Calcutta, India.

Singh, N. P. and S. Karthikeyan (2000) Flora of Maharashtra State: Dicotyledons Vol. I. Bot. Surv. India, Calcutta, India.

Singh, N. P. and S. Karthikeyan (2000) Flora of Maharashtra State: Dicotyledons Vol. II. Bot. Surv. India, Calcutta, India.

Ugemuge, N.R. (1986) Flora of Nagpur District. Shree Prakashan, Nagpur, India.

UNEP (1992) Convention on Biological Diversity, Rio de Janeiro.

*Wilt, D.O. (1911) List of fodder grasses found in the Derar forest circle of the central provinces. Pioneer Press Allahabad (1916) Descriptive list of trees, shrubs, climbers and economic herbs of the Northern and Berar Forest Circles. Central Provinces Allahabad.

Watt, George (1889–1896) A Dictionary of Economic Products of India, Vol. 1–6. Periodical Expert, Shahadara, Delhi, India.

Watt, George (1908) The Commercial Products of India: Being an Abridgement of the Dictionary of Economic Products of India. London.

* Original not consulted.

Index